E £10-95.

R L GARDNER Ph. D.

D1795809

The Cultured Cell and
Inherited Metabolic Disease

Previous Symposia of the Society for the
Study of Inborn Errors of Metabolism*

1. *Neurometabolic Disorders in Childhood. Ed. K. S. Holt and J. Milner 1963*
2. *Biochemical Approaches to Mental Handicap in Children. Ed. J. D. Allan and K. S. Holt 1964*
3. *Basic Concepts of Inborn Errors and Defects of Steroid Biosynthesis. Ed. K. S. Holt and D. N. Raine 1965*
4. *Some Recent Advances in Inborn Errors of Metabolism. Ed. K. S. Holt and V. P. Coffey 1966*
5. *Some Inherited Disorders of Brain and Muscle. Ed. J. D. Allan and D. N. Raine 1969*
6. *Enzymopenic Anaemias, Lysosomes and other papers. Ed. J. D. Allan, K. S. Holt, J. T. Ireland and R. J. Pollitt 1969*
7. *Errors of Phenylalanine Thyroxine and Testosterone Metabolism. Ed. W. Hamilton and F. P. Hudson 1970*
8. *Inherited Disorders of Sulphur Metabolism. Ed. N. A. J. Carson and D. N. Raine 1971*
9. *Organic Acidurias. Ed. J. Stern and C. Toothill 1972*
10. *Treatment of Inborn Errors of Metabolism. Ed. J. W. T. Seakins, R. A. Saunders and C. Toothill 1973*
11. *Inborn Errors of Skin, Hair and Connective Tissue. Ed. J. B. Holton and J. T. Ireland 1975*
12. *Inborn Errors of Calcium and Bone Metabolism. Ed. H. Bickel and J. Stern 1976*
13. *Medico-Social Management of Inherited Metabolic Disease. Ed. D. N. Raine 1977*

The Society exists to promote exchanges of ideas between workers in different disciplines who are interested in any aspect of inborn metabolic disorders. Particulars of the Society can be obtained from the Editors of this Symposium.

* Symposia 1–10 published by E. & S. Livingstone

The Cultured Cell and Inherited Metabolic Disease

MONOGRAPH BASED UPON
Proceedings of the Fourteenth Symposium of
The Society for the Study of Inborn Errors of Metabolism

EDITED BY

R. A. Harkness

and

F. Cockburn

MTP

Published by
MTP Press Limited
PO Box 55, St. Leonard's House
St. Leonardgate, Lancaster, England

© 1977 The Society for the Study of Inborn Errors of Metabolism

First Published 1977

No part of this book may be reproduced
in any form without permission from the
publisher except for the quotation of
brief passages for the purposes of review

ISBN 0 85200 167 3

Text set in 11/12 pt Photon Imprint,
printed by photolithography and bound in
Great Britain at The Pitman Press, Bath

Contents

Milner Lecture

Short Papers

Preface

The use of cultured cells in the clinical diagnosis of hereditary metabolic disease is a rapidly developing subject to which many different disciplines have brought their expertise and knowledge. A number of scientists who have individually contributed to the growth of the subject gave invited papers at the Fourteenth Symposium of the Society for the Study of Inborn Errors of Metabolism in the University of Edinburgh on 13–16th July, 1976. These papers form the basis of this monograph which brings together contributions from the basic sciences and from physicians concerned primarily with human disease. The cross-fertilization produced by this interdisciplinary communication was invaluable to those trying to understand and overcome diagnostic problems posed by hereditary metabolic disease.

Cell culture methods and cell preservation techniques were described by D. G. Harnden and D. E. Pegg; Dr T. Elsdale outlined some of the factors which control *in vitro* cell growth and division. Cell culture methods and cryopreservation techniques have allowed the wide distribution of biochemically abnormal cells and their study over long periods of time. It is also evident that when a defect which produces severe metabolic disorder in man can be studied in the laboratory using isolated cell cultures a wide variety of investigative procedures can be focused on to the cellular defect without distress or discomfort to the patient or relatives. Many different laboratories employing their own specialized techniques can contribute to the diagnosis, management, cure and prevention of the metabolic defect. Confirmation of results is needed because there are many pitfalls in work on cultured cells as shown by S. O. Lie in his contribution on the influence of cell culture medium on the metabolic behaviour of the fibroblast.

Cultured cells have so far been used extensively in studies on generally distributed systems such as the lysosomal acid hydrolases. In recent years the culture of specialized cells has been developing and is reviewed for neuromuscular systems by V. Dubowitz and for neurons and glial cells by E. J. Thompson. Both have suggested possible future applications for these surprisingly highly developed methods.

The biochemical section opens with an outline of the biochemistry of cytodifferentiation which shows many areas of our present knowledge with no apparent 'pathology' and aptly concludes with a plea to clinicians for evidence of such abnormalities. In the following comparison of leucocytes and cultured 'fibroblasts' in diagnosis, the difference is shown between a differentiated cell, the phagocytic microbicidal polymorphonuclear neutrophil leucocyte and the basic undifferentiated cultured 'fibroblast'. This section also contains a framework for the individual contributions to this monograph and an outline of present practice in the diagnosis of hereditary metabolic disease.

The next three contributions are concerned with deficiencies of lysosomal acid hydrolases which cause a wide range of storage disorders. The value of communication between the basic sciences and medicine is beautifully illustrated by these papers from Germany, Holland and Australia.

Defects in control of biosynthesis is considered next with a review of a defect in the control of cholesterol biosynthesis by a geneticist, F. W. Robertson.

Finally in the Milner Lecture, *Newer Developments in Tissue Culture: Further Aid in the Study of Inborn Errors of Metabolism*, Professor J. E. Seegmiller's own many contributions to our knowledge of the biochemistry of purines are reviewed. He also discusses the transformation of lymphocytes and the establishment of stable cell lines which carry defects. This exciting development may well further the progress of the study of inborn errors of metabolism, and of biochemistry. It should also help the many families who carry such defects.

Invited papers of the symposium were followed by 21 papers and demonstrations by members of the Society. The majority of papers presented new data which were a further but more detailed development of the theme of the meeting and fell into three categories, cultured cells and organic acid metabolism, cultured cells and acid hydrolase deficiencies and methodological developments. These were preceded by invited introductions by D. Gompertz on organic acid metabolism and by M. F. Niermeijer on methodological developments. These papers are listed in this monograph and will be published.

Our main aim has been to provide a practical and theoretical background for the very varied requirements of work on hereditary metabolic disease using cultured cells. We therefore hope that our attempt to provide a framework for interdisciplinary communication which has been so well supported by the authors themselves will be of help in medical practice, research and possibly in teaching.

We must express the thanks of all involved to Mr J. Milner for his generous support. We were privileged to have Dr G. Komrower as president, host and chairman during the meeting. It is also a pleasure to thank the City of Edinburgh for its warm and generous hospitality which was so much enjoyed by all, particularly by our many overseas colleagues. This meeting would not have been possible but for the kindness and efficiency of many members of the staff of the University of Edinburgh and of its teaching hospitals, especially the

Royal Hospital for Sick Children, Edinburgh and the staff of the Biochemistry Department at Alder Hey Children's Hospital. Our special thanks are also due to the many people who have helped in the preparation of this book.

Forrester Cockburn
R. Angus Harkness

List of Contributors

A. ADAMS
Department of Paediatric Biochemistry, Royal Hospital for Sick Children, Sciennes Road, Edinburgh EH9 1LJ

N. R. BELTON
Department of Child Life and Health, University of Edinburgh Medical School, Hatton Place, Edinburgh EH9 1UW

G. T. N. BESLEY
Department of Pathology, Royal Hospital for Sick Children, Sciennes Road, Edinburgh EH9 1LF

M. E. BLASCOVICS
Children's Hospital, 4650 Sunset Boulevard, Los Angeles, California 90027, USA

K. B. BLAU
Barron Memorial Research Laboratories, Queen Charlotte's Maternity Hospital, Goldhawk Road, London W6 OXG

N. J. BRANDT
Department of Teratology, Rigshospitalet, Juliane Mariesvej, DK2100 Copenhagen, Denmark

D. M. BROADHEAD
Department of Pathology, Royal Hospital for Sick Children, Sciennes Road, Edinburgh EH9 1LF

EDNA BROWN
Division of Inherited Metabolic Disease, MRC Clinical Research Centre, Northwick Park Hospital, Watford Road, Harrow, Middlesex

J. BUTTERWORTH
Department of Pathology, Royal Hospital for Sick Children, Sciennes Road, Edinburgh EH9 1LF

NINA CARSON
Nuffield Department of Child Health, Institute of Clinical Science, Grosvenor Road, Belfast BT12 6BJ, Northern Ireland

R. A. CHALMERS
Division of Inherited Metabolic Disease, MRC Clinical Research Centre, Northwick Park Hospital, Watford Road, Harrow, Middlesex

E. CHRISTENSEN
Department of Teratology, Rigshospitalet, Juliane Mariesvej, DK2100 Copenhagen, Denmark

F. COCKBURN
Paediatric Laboratory, Simpson Memorial Maternity Pavilion, Lauriston Place, Edinburgh EH3 9EF

V. DUBOWITZ
Department of Paediatrics, Royal Postgraduate Medical School, Hammersmith Hospital, Du Cane Road, London W12 OHS

M. DURAN
Wilhelmina Kinderziekenhuis, Nieuwe Gracht, Utrecht, The Netherlands

T. ELSDALE
MRC Clinical Population and Cytogenetics Unit, Western General Hospital, Crewe Road, Edinburgh 4

A. E. H. EMERY
Department of Human Genetics, Western General Hospital, Crewe Road, Edinburgh 4

J. W. FARQUHAR
Department of Child Life and Health, University of Edinburgh Medical School, Hatton Place, Edinburgh EH9 1UW

K. von FIGURA
Physiological Chemistry Institute, University of Münster, Waldeyer Strasse, D4400 Münster, West Germany

B. FOWLER
Department of Human Genetics, Yale University School of Medicine, New Haven, Connecticut 06510, USA

M. J. FRANCIS
Nuffield Orthopaedic Centre, Headington, Oxford OX3 7LD

H. GALJAARD
Department of Cell Biology and Genetics, Medical Faculty, Erasmus University, Rotterdam, The Netherlands

R. GITZELMANN
Division of Metabolism, University Paediatric Department, Kinderspital, CH8032 Zurich, Switzerland

D. GOMPERTZ
Royal Postgraduate Medical School, Hammersmith Hospital, Du Cane Road, London W12 0HS

D. G. HARNDEN
Department of Cancer Studies, University of Birmingham Medical School, Birmingham B15 2TJ

R. A. HARKNESS
Division of Perinatal Medicine, MRC Clinical Research Centre, Northwick Park Hospital, Watford Road, Harrow, Middlesex

J. HOLTON
Department of Chemical Pathology, Southmead Hospital, Westbury-on-Trym, Bristol BS10 5NB

R. C. HUGHES
National Institute for Medical Research, Mill Hill, London NW7 1AA

J. T. IRELAND
Department of Biochemistry, Alder Hey Children's Hospital, West Derby, Liverpool L12 2AP

G. KOMROWER
The Park Hospital for Sick Children, Old Road, Headington, Oxford OX3 7LQ

S. O. LIE
Paediatric Research Institute, University of Oslo, Rigshospitalet, Oslo 1, Norway

MONIQUE MATHIEU
Department of Biochemistry, Hopital Debrousse, Rue de Soeur Bouvier, F69322 Lyon Cedex 1, France

IRENE MARIE
Department of Biochemistry, Hopital Debrousse, Rue de Soeur Bouvier, F69322 Lyon Cedex 1, France

J. MILNER
Menlove Gardens, Liverpool 18

P. MILNER
Menlove Gardens, Liverpool 18

M. F. NIERMEIJER
Department of Cell Biology and Genetics, Erasmus University, Rotterdam, The Netherlands

D. E. PEGG
Department of Cryobiology, MRC Clinical Research Centre, Northwick Park Hospital, Watford Road, Harrow, Middlesex

J. M. RATTENBURY
Barron Memorial Research Laboratories, Queen Charlotte's Maternity Hospital, Goldhawk Road, London W6 0XG

A. J. J. REUSER
Department of Cell Biology and Genetics, Medical Faculty Erasmus University, Rotterdam, The Netherlands

F. W. ROBERTSON
Department of Genetics, Marischal College, University of Aberdeen, Aberdeen AB9 1AS

A. B. ROY
Department of Physical Biochemistry, John Curtin School of Medical Research, Australian National University, Canberra, A.C.T. 2601, Australia

I. B. SARDHARWALLA
Willink Biochemical Genetics Unit, Royal Manchester Children's Hospital, Pendlebury, Manchester M27 1HA

J. E. SEEGMILLER
Department of Medicine M-103, School of Medicine, University of California, La Jolla, California, 92003, USA

W. M. TELLER
Department of Paediatrics, University of Ulm, Prittwitzste, D7900 Ulm, West Germany

E. J. THOMPSON
Department of Neurochemistry, Institute of Neurology, Queen Square, London WC1 3BG

C. TOOTHILL
Department of Chemical Pathology, University of Leeds Medical School, Leeds, LS2 9NL

D. E. S. TRUMAN
Department of Animal Genetics, University of Edinburgh, Edinburgh EH8 9YL

S. K. WADMAN
Wilhelmina Kinderziekenhuis, Nieuwe Gracht, Utrecht, The Netherlands

R. W. E. WATTS
Division of Inherited Metabolic Disease, MRC Clinical Research Centre, Northwick Park Hospital, Watford Road, Harrow, Middlesex

Cell Biology

1
Cell Biology and Cell Culture Methods—A Review

D. G. HARNDEN

Department of Cancer Studies, University of Birmingham

The study of inborn errors of metabolism dates from Garrod's classical studies in alcaptonuria published in 1908. His work was based on straightforward chemical analysis combined with careful observation of the family structure. Similarly, the origins of cell culture can be traced back to the experiments of Wilhelm Roux on chick embryos in 1885, but it is generally accepted that Ross Harrison's experiments with the culture of frog nervous tissue in clots of frog lymph, carried out in 1907, are the first true tissue culture experiments in the sense we accept today. It is curious that these fundamental studies should have originated about the same time and yet the impact of the one upon the other should not have been felt for more than half a century. This is not the place to review the development of the various techniques of tissue culture but it is worth noting that these were developed to a large extent with frog, chick and mouse tissue. The culture of normal human cells was developed at a much later date. Human tumour cells were, however, being cultured successfully by Losee and Ebelling as early as 1914 and by 1936 Gey was culturing human tumours on a considerable scale. It was at that time Gey also showed that the culture of normal human cells from surgical specimens was not as difficult as previously suggested. There was nevertheless a long gap thereafter in the development of normal human cell culture techniques. Many people found that cell lines like Hep1 and HeLa which appeared to grow indefinitely in culture were more convenient to use, and the degree to which these cells were abnormal was not fully appreciated at that time. In 1956 the use of primary cells received considerable stimulus from the recognition by Tjio and Levan, using human embryo cell cultures, that the normal human chromosome number was 46.

Shortly afterwards Puck *et al.* (1958) showed that it was possible to grow cell cultures from small skin biopsies by trypsinizing the tissue and seeding out the cells. At about this time, and inspired largely by the work of Puck, I developed a technique for the growth of fibroblast cultures from skin biopsies using a modification of the plasma clot explant technique which had been in use for many years for other tissues (Harnden, 1960). The advantage of this technique was that it gave good results in relatively inexpert hands. At first human skin fibroblast cultures were used largely for cytogenetic studies, but their use for other studies soon became widespread. In 1960, Krooth and Weinberg found that fibroblasts from patients with galactosaemia failed to grow in medium with galactose as the only energy source. Other demonstrations of inborn errors in culture soon followed, e.g. glucose-6-phosphate dehydrogenase deficiency (Gartler et al., 1962) and acatalasia I (Krooth et al., 1962). The next major step was the demonstration that not only could exfoliative cytology be carried out on amniotic fluid cells but that, amongst these cells, was a population of cells which could grow rapidly and from which chromosome preparations could be made (Steele and Breg, 1966). Antenatal diagnoses of inborn errors such as mucopolysaccharidosis (Nadler, 1968) followed soon thereafter as did recognition of some heteroxygotes, e.g. Lesch–Nyhan syndrome (Fujimoto et al., 1968). Syndromes which demonstrate indirectly an inborn error such as the UV sensitivity of xeroderma pigmentosum (Cleaver, 1968) and the X-ray sensitivity of ataxia telangiectasia (Taylor et al., 1975) can now also be recognized in cell culture.

For more extensive reviews on this subject the reader is referred to the books of Paul (1975) and Littlefield (1976).

METHODS OF CULTURE

Only culture methods relevant to the study of inborn errors of metabolism will be considered. Details of the techniques are given elsewhere (Yunis, 1974; Paul, 1975).

Organ culture

Unlike cell culture the aim of organ culture is to preserve in vitro the main architecture of the tissue from which the biopsy was obtained. There is a variety of methods but all aim to hold the tissue immobile in a liquid medium under conditions of high or relatively high oxygenation. In Trowell's (1954) technique the tissue is chopped very carefully into small pieces and placed on a piece of lens paper on a stainless steel grid which supports the tissue at the surface of the nutrient medium. The culture dish is placed in a special chamber in which the gas phase can be controlled at 5% CO_2 in oxygen. Using such a system the tissue can be kept in good condition for about seven days. Although there would be advantages in such a technique for studying certain inborn errors, in particular those diseases affecting the intercellular matrix, it seems to have been remarkably little used, e.g. in the excellent recent monograph of the Society for Cell Biology on 'Organ Culture in Biomedical Research' the subject is not mentioned (Balls and Monnickendam, 1975).

Cell cultures

It is possible to establish cell cultures from tissue samples of almost any tissue. Some, like liver, are particularly difficult to grow. In others a heterogeneous population of cells will grow, often a mixture of epithelial and fibroblastic cells. Fibroblastic cells always come to predominate. Almost pure epithelial cell cultures can, however, be obtained from some tissues, e.g. kidney. Such tissue samples can usually only be obtained during major or minor surgery and therefore are of value only on special occasions. Similarly, tissue taken at autopsy can be grown but again this is not generally applicable.

Emphasis, therefore, will be placed firmly on the use of cells derived from skin biopsies, from blood and from amniotic fluid since they can be obtained routinely from any subject without major risk or discomfort.

Fibroblast cultures from skin biopsies

There are a variety of different procedures described in detail elsewhere (e.g. Harnden, 1974). Using a local anaesthetic the biopsy may be taken by means of a punch or simply by lifting the skin on the top of a needle and cutting off a tiny portion with a scalpel blade. The method preferred by us is to pick up a small fold of skin (perhaps 1–2 mm × 3–4 mm) using fine forceps and, after allowing the skin to become numb, cutting the fold off with a scalpel blade. Anaesthetic is not necessary for this procedure. This is a good method for children who usually seem quite untroubled by the procedure. It tends to give a rather small biopsy and there is some relationship between the size of the biopsy and the time taken to the first subculture. The tissue is placed in complete growth medium in a screw-capped bottle and transferred to the tissue culture laboratory. It can be sent long distances, remaining in the bottle for at least five days, and still give good growth. The culture is established by cutting the biopsy into small pieces and anchoring these in some way in the culture vessels (always at least two dishes in parallel). Currently, we simply place a sterile microscope cover slip over the tissue and then add Ham's F10 medium with 20% fetal calf serum. Normally, we add penicillin (100 units/ml) and streptomycin (100 μg/ml) but after the culture is established this can be omitted.

The outgrowth may show epithelial cells at first but soon fibroblasts grow out and, on subculture, totally replace the epithelial cells. Cultures are fed regularly with fresh medium even when only a small outgrowth is present. The growth under a cover slip is not as fast as in plasma clots with embryo extract (which I used originally) but possible contamination with chicken viruses or even cells is avoided. The first subculture is usually made after about two weeks but in the case of very small biopsies or biopsies from some abnormal patients it may be longer. Some cells always adhere to the cover slip, which can be removed and used for rapid preliminary tests or to seed out a new culture. Subcultures are made every 4–7 days by removing the cells from the culture vessel with a mixture of trypsin and EDTA and diluting the cells 1 to 2 into fresh medium. As Hayflick (1965) has pointed out primary human cells cannot be cultured indefinitely but will eventually die out after about 50–60 doublings for embryo fibroblasts and about 30–40 doublings for adult skin fibroblasts. Some metabolically abnormal cells will not grow for even this period. In our experience, 20 generations is good for cells from patients with mucopolysaccharidoses while in the case of ataxia telangiectasia cells from some patients grow as well as do normal cells while others are hard to grow at all. However, even with cells from abnormal subjects, relatively large numbers of cells can usually be grown.

This technique can be used in exactly the same way for any other tissue specimen but for embryo cells it is usually sufficient simply to chop the tissue (kidney, lung, skin) finely and place it in a culture dish with medium. If left undisturbed for 2–3 days this will rapidly give confluent cell cultures.

The choice of vessels for culture depends to a large extent on the use to which the cells will be put. Routinely we use 5 cm Nunc plastic (TC grade) petri dishes for maintenance of cell stocks. However there is a wide range of satisfactory plastic or glass dishes and bottles.

Cells may be grown on a small or large scale. For many purposes, for example cell survival curves to measure response to an environmental agent or to measure mutation frequency in a selective system, cells will have to be grown from single cells, i.e. cloned. This is usually done in 10 cm dishes using a feeder layer of 10^5 cells of the same type irradiated with 10 K rad of X-rays. The cells to be grown are plated out in small numbers with the irradiated cells. For cloning we use Eagle's MEM medium with 20% of fetal calf serum and obtain about 30% plating efficiency. Again, some metabolically abnormal cells give consistently lower plating efficiencies. We would aim to get about 50 colonies per dish since a greater number makes counting difficult.

For large scale cultures the cells are seeded into Winchester size roller bottles with about 100 ml medium. The bottles are rolled gently on a motor driven device at about one turn every eight minutes. The cells come to coat the entire inside surface of the bottle. This technique, which is very successful for cell lines, does not always give high yields of adult fibroblasts. It does, however, seem excellent for human embryo kidney cells and we regularly produce our stocks of adenovirus type 2 on such cultures.

Amniotic fluid cell cultures

A number of techniques, some complex and some very simple, have been devised. In our experience we find it important to set up the cultures as soon as possible after amniocentesis and to handle the fluid as little as possible. We place about 3–4 ml of amniotic fluid and an equal volume of culture medium containing 20% fetal calf serum in a plastic petri dish containing a sterile glass coverslip. At least two dishes are set up from each fluid, incubated and left undisturbed for 3 or 4 days. The medium is then changed and clonal growth of the cells is observed after a further 3 or 4 days. After about 2–3 weeks on average, the cells are ready for subculture or for rapid studies on any clones which have settled down on the cover slips. The examination of individual clones also has the advantage of helping to identify contamination with maternal cells. Other groups (e.g. Nadler and Ryan, 1974) have found that spinning down the cells and transferring them to a culture vessel in which they are immobilized with a cover slip before the medium is added, gives good results. The handling of the cells is then the same as for fibroblast cultures. Some very elegant microtechniques (Galjaard et al., 1974) have been devised and in expert hands these have proved to be an accurate and very rapid method of carrying out cytochemical tests on cultured cells.

Blood culture

Short term—For certain types of study short term blood cultures are useful. These are based on the method of Moorhead et al. (1960) which in turn was based on the work of Osgood and Krippaehne (1955). Though these cultures are more often used for cytogenetic or immunological studies they are so simple to carry out that their application is wide. Either whole blood or purified lymphocytes (Zackai and Mellman, 1974) can be mixed with growth medium containing 10% bovine serum and phytohaemagglutinin which is a mitotic stimulant for lymphoid cells of the T series. Other mitogens can be used (e.g. pokeweed or concanavalin A). After 48–72 h large numbers of cells are in mitosis. One advantage of this system is that the cells are all in the G_0 phase of the cell cycle at the initiation of the culture and the first wave of mitosis is to some extent synchronous. The cells will not continue to divide for more than two or three cycles which limits the usefulness of this method.

Long term—Continuous cultures of lymphoid cells may be established either by cultivating normal lymphoid cells with irradiated cells from other established lymphoid cell lines which contain EB virus genome or by exposing them to high concentrations of the virus itself. In this case it is the B cell series which grows out and it now seems firmly established that the presence of the EB virus genome is essential for the long term cultivation of 'B lymphocytes'. They are in fact virus-transformed cells. Provided that this is borne in mind these cultures clearly have their uses in the study of inborn errors. Povey et al. (1973) have studied isoenzyme variants of a range of such cell lines derived from different individuals. In the past one handicap has been the difficulty of establishing cell lines regularly from any individual. There are now a number of laboratories which can reproducibly create a lymphoid cell line from any sample of peripheral blood (Dr A. Greene, personal communication). Since these cells grow in suspension they can be grown in bulk in spinner cultures and enormous quantities of cells produced if required. Essentially a large glass vessel is seeded with cells in a specially designed suspension medium (usually low in calcium). Cells are kept in suspension with a magnetic stirrer and medium added as the cell numbers increase. A number of automatic systems for addition of medium and removal of cells has been devised. Attempts to grow normal human fibroblasts in such systems have not to my knowledge been successful. However, other established cell lines have been adapted to grow in this manner. In addition, some cell types will grow in suspension in soft (0.3%) agar. While normal cells will not divide under such conditions, transformed cells (Macpherson and Montagnier, 1964) and some haemopoietic stem cells (Metcalf *et al.*, 1967) will grow to form quite large colonies. Lymphoid cell lines will grow in soft agar but with only a low plating efficiency.

PROBLEMS

Alterations in metabolism

One of the criticisms of all *in vitro* work is that the cells do not truly reflect the behaviour of cells *in vivo*. This fear is particularly acute in the study of inborn errors of metabolism. Cells in culture are often said to dedifferentiate and to some extent this is true in that they are made to proliferate rapidly. They acquire an unusually high level of glycolysis and there is some evidence of a switch in enzyme function. For example, Owens and Nebert (1974) have shown that, in a wide range of cultured tissues, the cytochrome P_{450} mixed function oxidase is hard to activate while the cytochrome P_{448} oxidase which is characteristic of more rapidly dividing cells seems to take over its function. In particular, drugs such as phenobarbitone which normally induce cytochrome P_{450} *in vivo* induce cytochrome P_{448} in cultured cells. On the other hand the fact that an enormous range of metabolic defects are expressed in cell culture suggests that, most often, mutant enzyme patterns will be truly reflected in cell culture.

One further problem which has afflicted our laboratory from time to time is the appearance of 'blobs' of material in the cytoplasm of the cells which will stain with orcein and a number of other stains. This material has not been properly identified but may contain a mucopolysaccharide. It occurs only in cells cultured in calf serum—such cells when transferred to human serum lose these abnormal bodies within 2–3 days. We take this to be a morphological manifestation of abnormal metabolism in these cells. These bodies are now appearing in pictures of cultured cells from a number of other laboratories and should be more thoroughly investigated.

Physical problems

Maintenance of pH, temperature and osmolarity are all critical for good cell culture and are discussed in more detail by Lie in this volume.

Contamination

In practice, much more important than alterations in metabolism is the problem of contamination.

Microorganisms—The best safeguard of all is good technique on the part of the operator. Nevertheless, contamination of cultures with yeasts and fungi happens from time to time. We do not consider it worthwhile to treat the cells and we always discard them. Bacterial contamination, if it occurs, can often be controlled by antibiotics long enough to make critical observations, but it is wise to adopt some sort of antibiotic policy. While it is convenient to grow cultures routinely in penicillin and streptomycin, it is preferable after establishment of the cultures to switch any cells that are to be stored in a cell bank into an antibiotic-free medium so that it can be certain that banked cells are clean.

More troublesome than bacterial contamination is mycoplasma, since a variety of different strains may grow in cultured cells. Any laboratory working with cell culture should have some routine procedure for mycoplasma testing. It is essential to emphasize this when cells are being used for metabolic studies since, although the biochemistry of the mycoplasmas may be interesting it is important to keep it separate from the biochemistry of the cells. Deliberate culture of mycoplasmas is slow and not easy to do routinely. Mycoplasmas can be simply detected by autoradiography or fluorescence microscopy. In the latter technique (Russell *et al.*, 1975) cells are stained with a fluorescent dye such as Hoechst 33258 and examined under a UV microscope. The mycoplasmas can be clearly seen as fluorescent bodies coating the surface of the cells. There is no simple way of keeping cultures free from mycoplasmas although avoiding mouth pipetting appears to help. Routine checks should lead to rapid detection and avoid confusion. We rarely see contaminations in primary cells or indeed cultures of human diploid cells. Any contaminated cultures are discarded immediately.

Contamination with virus is relatively easy to detect since usually it will destroy the culture but this is a rare occurrence even in a laboratory such as ours which is handling viruses. A transforming virus such as SV 40 might remain undetected for longer but the abnormal behaviour of the cells would again betray its presence which could be checked simply by looking for T antigen. Such an accidental contamination has never occurred in our laboratory although we handle SV 40. Some concern that cell cultures might harbour C-type viruses similar to members of the oncorna virus group has recently been expressed. These would not be apparent since the virus is shed from the cell without killing it. Particular concern has been felt about, firstly, lines established from human tumours especially when first grown in nude mice and, secondly, hybrid cells where one parent is human and the other another mammalian species (see below). The extent of this hazard is not known and should be kept under review.

Cells—Contamination with other cells has been recognized for a long time. Many of the early so-called spontaneous transformations proved to be contaminations with cells of a different type. However, the problem is still with us. Apart from wrong labelling which can be avoided and the occasional growth of maternal cells in amniotic fluid cell cultures, the main problem is still accidental contamination with established cell lines. Gartler (1968) has shown using the glucose-6-phosphate dehydrogenase and phosphoglucomutase phenotypes that a high proportion of established cultures are in fact HeLa cells. There should be no confusion where the line being studied is a diploid cell line since there is a distinct morphological difference. Checks can be made of both isoenzyme variants and of cytogenetics to ensure that a cell culture is indeed what it is supposed to be. Fortunately, cells from patients with inborn errors have built-in markers.

HAZARDS

Hepatitis

All human blood and tissue should be treated with care and in our laboratory all such samples are treated as if virus-contaminated.

Shedding of viruses

Apart from the possible shedding of C-type virus mentioned above, lymphoid cell lines may be shedding EB virus. At present the degree of hazard is unknown but it is again normal procedure to treat all such cultures as potentially virus-contaminated.

Exploding ampoules

When cells are stored in sealed glass ampoules immersed in liquid nitrogen there is always a danger that an explosion may occur when the ampoule is removed from the cell bank. For this reason face masks should always be worn. Risk may be minimized if the sealed ampoules are, before freezing, first placed in a chilled jar of alcohol containing a brightly coloured dye (Dr A. Greene, personal communication). If the ampoule has a hole in it, dye is sucked in and the coloured ampoule may be discarded. Some workers believe that cells should always be stored in vapour phase liquid nitrogen refrigerators.

SPECIAL TECHNIQUES

A variety of the special techniques which are being applied to cell culture will be dealt with in detail by other speakers in this symposium, but one or two can be mentioned here.

Cell banking

The storage of cells in liquid nitrogen is now standard procedure in many laboratories. Dr Pegg will deal with the details but I should like to draw attention to two practical problems. First, in the USA, a cell bank for human cell types has been established at the Institute for Medical Research at Camden, New Jersey. We should consider now whether or not such a facility should be set up in Western Europe. It would obviously be a very expensive exercise both to establish and to run. If we are not to do this we must take steps to ensure that laboratories storing cells maintain certain standards in quality of the stocks and in record keeping. To do this will cost money and I am suggesting that a specific central fund should be established by the Department of Health and Social Security and the Medical Research Council to support the accurate banking of human cell stocks. In comparison the second problem is trivial but none the less real. Scientists are very happy to store cells but they will not discard them. Our cell bank has become a sort of monster that grows a new head every year. This is a problem of laboratory management but it must be solved rationally without discarding valuable cell stocks.

Mutation

The classical studies of Kao and Puck (1968) demonstrated that mammalian cells could be used in mutation studies in a way very similar to microorganisms. They demonstrated that cells with a stable mutant phenotype could be selected out using cloning techniques in media which exerted selective pressures by blocking off specific metabolic pathways. So far studies with human cells have proved difficult technically but several groups have had considerable success. In particular, De Mars and his colleagues have carried out highly sophisticated studies (e.g. Rappaport and de Mars, 1973). Cells from patients with some metabolic disorders have proved useful in these studies since they provide a ready made selective system, e.g. Lesch–Nyhan cells which are defective in HGPRT will not grow in a medium containing aminopterin even with added sources of hypoxanthine and thymidine (HAT) whereas normal cells will grow quite well in such a medium. Mutant lines with such mutant phenotypes can also be derived from normal cells by exposure to chemical mutagens and growth in a selective medium. However, with normal human adult cell cultures the cells go through so many cell doublings in the process of cloning that even though such mutants can be selected they are of limited use once the clones have been chosen.

Other types of metabolically abnormal cell that will be of particular value are those from patients with xeroderma pigmentosum (XP) which are unusually sensitive to UV and those from patients with ataxia telangiectasia (AT) which are unusually sensitive to X-rays.

Cellular hybridization

There is little need to stress the importance of cellular hybridization in the study of inborn errors. Cells may fuse spontaneously (Barski *et al.*, 1960), but more efficient fusion can be promoted by adding a virus which normally causes syncytium formation in the course of infection. The virus normally used, Sendai virus, is first inactivated with UV light or treatment with β propiolactone. The two parental cell suspensions at about 10^7 cells/ml are mixed with the inactivated virus and held at $4\,^{\circ}C$ for 15 minutes. Clumped cells are then plated out in growth media at $37\,^{\circ}C$. Various selective systems have been devised for getting rid of parental cells and homokaryons, e.g. an $HGPRT^-TK^+$ parent line fused with an $HGPRT^+TK^-$ line would yield $HGPRT^+TK^+$ heterokaryons that could grow in HAT selective medium, whereas the parental lines and any homokaryons would not survive. Other features such as sensitivity to high temperatures or failure to grow in calcium-containing media have also been used as part of the selective system. Such hybridization techniques are playing an important role in studying chromosomal localization of specific genes and in linkage studies which will be vitally important in antenatal diagnosis.

Cytogenetics

Chromosome studies are dealt with extensively elsewhere. Their use in the study of inborn errors is essentially a negative one. If an abnormal child is shown to have a chromosomal abnormality there is less point in looking for a specific enzyme defect. Some translocations and deletions have however been shown to be associated with specific metabolic changes (e.g. deletion of the short arms of chromosome 18 with deficient IgA).

FUTURE

We have only just begun to see the possibilities of the culture of human cells but already techniques of cell culture have contributed much to the understanding of human metabolic disease. In the future I see two main fields of activity, firstly, the utilization of cells with metabolic defects to promote studies in both fundamental and applied biology and secondly the further application of cell culture techniques to the management of patients with metabolic defects.

As an example of the first category, I should like to consider short-term testing for mutagenesis and carcinogenesis. A great variety of tests are available. The bacterial tests of Ames *et al.* (1973), Garner *et al.* (1975) and others are so easy to carry out that they clearly have a place in such work. Similarly, animal testing will continue to have its place but animal tests are time-consuming and expensive and like the bacterial tests their relevance to the human situation may be questioned. The use of human mutation systems utilizing cells from metabolically abnormal patients may come to play a vital role in the screening of potentially hazardous chemicals. Ames' strains of *Salmonella typhimurium* are particularly suitable for mutagenicity testing because they are DNA repair deficient and cell wall deficient. Already we have two human cell types that are DNA repair deficient (AT and XP) and mutation systems using these cells have been developed. It is not too hard to imagine the construction of a supersensitive human hybrid cell that would become the standard system for the testing of carcinogens and mutagens.

In the application of cell culture to the study of inborn errors we will obviously have further development of tests for more and more syndromes. But we must be careful to avoid collecting new tests like postage stamps especially when the conditions in whose diagnosis they are intended to help are so rare that the chances of a particular laboratory seeing even one case a year is remote. Attention must, therefore, be given to a rationalization of cell culture testing. We should acknowledge that the cells are easy to grow and store but that tests are often hard to carry out unless a considerable body of expertise is built up. The diagnostic function is, therefore, best placed not in the laboratories that grow the cells or in central biochemistry laboratories but in laboratories that have a research interest in the condition in question—this may be only one laboratory in, say, the UK or even Western Europe for each condition.

Developments in our knowledge of linkages will clearly extend our expertise in antenatal diagnosis faster than our development of new tests, and the creation of a new map for human chromosomes is progressing steadily. Another area on which attention should be firmly focused in the future is therapy. When a condition is known to be due to a defect of a specific enzyme rather than of a whole chromosome the possibility of treatment becomes more realistic. Cell culture may have an important role to play here, and cultures may eventually be used for the production of specific enzymes. The work of Dean *et al.* (1976) however, seems to point the way towards the use of the cells themselves for therapeutic purposes. They found by implanting subcutaneously a suspension of 2.2×10^8 cells from the normal sister of a patient with Hunter's syndrome that, not only was there biochemical evidence of an effect of the implanted cells upon metabolism of the patient, but also the progressive intellectual deterioration which is characteristic of these patients was at least held in check for a period of ten months and there was a general improvement in behaviour. Obviously this is only a pilot study but it should encourage others to investigate this intriguing area further.

References

Ames, B. N., Durston, W. E., Yamasaki, E. and Lee, F. D. (1973). Carcinogens are mutagens: A simple test system combining liver homogenates for activation. *Proc. Nat. Acad. Sci. U.S.A.,* **70,** 2281.

Balls, M. and Monnickendam, M. (1975). Organ culture in biomedical research. *British Society for Cell Biology Symposium,* **1.** (Cambridge University Press)

Barski, G., Sorieul, S. and Cornefert, F. (1960). Production dans de cultures *in vitro* de deux souches cellulaires en association de cellules de caractere 'hybrid'. *C. R. Acad. Sci. Paris,* **251,** 1825

Cleaver, J. E. (1968). Defective repair replication of DNA in xeroderma pigmentosium. *Nature (London),* **218,** 652

Dean, M. F., Muir, H., Benson, P. F., Button, L. R., Boylston, A. and Mowbray, J. (1976). Enzyme replacement therapy by fibroblast transplantation in a case of Hunter's syndrome. *Nature (London),* **261,** 323

Fujimoto, W. Y., Seegmiller, J. E., Uhlendorf, B. W. and Jacobson, C. B. (1968). Biochemical diagnosis of an X-linked disease *in utero. Lancet,* **ii,** 511

Galjaard, H., Van Hoogstraten, J. J., de Josselin de Jong, J. E. and Mulder, M. P. (1974). Methodology of the quantitative cytochemical analysis of single or small numbers of cultured cells. *Histochem. J.,* **6,** 409

Garner, R. C., Walpole, A. L., Rose, F. L., (1975). Testing of some benzidine analogues for microsomal activation to bacterial mutagens, *Cancer Lett.,* **1,** 39.

Garrod, A. E. (1908). The Croonian lectures on inborn errors of metabolism *Lancet,* **ii,** 1, 73, 142 and 214

Gartler, S. M. (1968). Apparent HeLa cell contamination of human heteroploid cell lines. *Nature (London),* **217,** 750

Gartler, S. M., Gandini, E. and Cepellini, R. (1962). Glucose-6-phosphate dehydrogenase deficient mutant in human cell culture. *Nature (London),* **193,** 602

Gey, G. O. and Gey, M. K. (1936). The maintenance of human normal cells and tumour cells in continuous culture. 1. Preliminary report: cultivation of mesoblastic tumours and normal tissue and notes on methods of cultivation. *Am. J. Cancer,* **27,** 45

Harnden, D. G. (1960). A human skin culture technique used for cytological examinations. *Br. J. Exp. Pathol.,* **41,** 31

Harden, D. G. (1974). Skin culture and solid tumour technique. In: *Human Chromosome Methodology,* J. J. Yunis (ed.), New York: Academic Press, p. 167

Harrison, R. G. (1907). Observations on the living developing nerve fibre. *Proc. Soc. Exp. Biol. Med.,* **4,** 140

Hayflick, L. (1965). The limited *in vitro* lifetime of human diploid cell strains. *Exp. Cell Res.,* **37,** 614

Kao, F. T. and Puck, T. T. (1968). Genetics of somatic mammalian cells. VII. Induction and isolation of nutritional mutants in Chinese hamster cells. *Proc. Nat. Acad. Sci. U.S.A.,* **60,** 1275

Krooth, R. S. and Weinberg, A. N. (1960). Properties of galactosemic cells in culture. *Biochem. Biophys. Res. Commun.,* **3,** 518

Krooth, R. S., Howell, R. R. and Hamilton, H. B. (1962). Properties of acatalasic cells growing *in vivo. J. Exp. Med.,* **115,** 313

Littlefield, J. W. (1976). *Variation, senescence and neoplasia in cultured somatic cells.* (Harvard University Press)

Losee, J. R. and Ebelling, A. H. (1914). The cultivation of human sarcomatous tissue *in vitro. J. Exp. Med.,* **20,** 140

Macpherson, I. A. and Montagnier, L. (1964). Agar suspension culture for the selective assay of cells transformed by polyoma virus. *Virology,* **23,** 291

Metcalf, D., Bradley, T. R. and Robinson, W. (1967). Analysis of clones developing *in vitro* from mouse bone marrow cells stimulated by kidney feeder layers and leukaemic serum. *J. Cell. Physiol.*, **69,** 93

Moorhead, P. S., Nowell, P. C., Mellman, W. J., Battips, D. M. and Hungerford, D. A. (1960). Chromosome preparation of leukocytes cultured from human peripheral blood. *Exp. Cell Res.*, **20,** 613

Nadler, H. L. (1968). Antenatal detection of hereditary disorders. *Pediatrics*, **42,** 912

Nadler, H. L. and Ryan, C. A. (1974). Amniotic cell culture. In: *Human Chromosome Methodology*, J. J. Yunis (ed.)., p. 185 (New York and London: Academic Press)

Osgood, E. E. and Krippaehne, M. L. (1955). The gradient tissue culture method. *Exp. Cell Res.*, **9,** 116

Owens, I. S. and Nebert, D. W. (1975). Aryl hydrocarbon hydroxylase induction in mammalian liver-derived cell cultures. Stimulation of 'cytochrome $P1_{-450}$ associated' enzyme activity by many inducing compounds. *Mol. Pharmacol.*, **11,** 94

Paul, J. (1975). *Cell and tissue culture.* 5th Edition. (Edinburgh and London: Churchill Livingstone)

Povey, S., Gardiner, S. E., Watson, B., Mowbray C., Harris, H., Arthur, E., Steel, C. M., Blenkinsop, C. and Evans, H. J. (1973). Genetic studies on human lymphoblastoid lines: isoenzyme analysis on cell lines from forty-one different individuals and on mutants produced following exposure to a chemical mutagen. *Ann. Hum. Genet.*, **36,** 247

Puck, T. T., Cieciura, S. J. and Robinson, A. (1958). Genetics of somatic mammalian cells. III. Long-term cultivation of euploid cells from human and animal subjects. *J. Exp. Med.*, **108,** 945

Rappaport, H. and de Mars R. (1973). Diaminopurine-resistant mutants of cultured diploid fibroblasts. *Genetics*, **75,** 335

Roux, W. (1885). Beiträge zur Entwicklungsmechanik des Embryo. *Z. Biol.*, **21,** 411

Russell, W. C., Newman, C. and Williamson, D. H. (1975). A simple cytochemical technique for demonstration of DNA in cells infected with mycoplasmas and viruses. *Nature (London)*, **253,** 461

Steele, M. W. and Breg, W. R. (1966). Chromosome analysis of human amniotic fluid cells. *Lancet*, **i,** 383

Taylor, A. M. R., Harnden, D. G., Arlett, C. F., Harcourt, S. A., Lehman, A. R., Stevens, S. and Bridges, B. A. (1975). Ataxia telangiectasia: a human mutation with abnormal radiation sensitivity. *Nature (London)*, **258,** 427

Tjio, J. H. and Levan, A. (1956). The chromosome number of man. *Hereditas*, **42,** 1

Trowell, O. A. (1954). A modified technique for organ culture *in vitro*. *Exp. Cell Res.*, **6,** 246

Yunis, J. J. (1974). *Human Chromosome Methodology.* 2nd Edition. (New York: Academic Press)

Zackai, E. H. and Mellman, W. J. (1974). Human peripheral blood leucocyte cultures. In *Human Chromosome Methodology*, J. J. Yunis (ed.) 2nd Ed., p. 95. (New York and London: Academic Press)

2

Influence of Cell Culture Medium on the Metabolic Behaviour of the Fibroblast

S. O. LIE*

Paediatric Research Institute, National Hospital of Norway, Rikshospitalet, Oslo, Norway

Monolayer cultures of human fibroblasts are now well established tools for the study of inborn errors of metabolism in man. Indeed, one genetic disease (I-cell disease) derives its name from the inclusions found in cultured cells from two children (Leroy and DeMars, 1967). However, it is highly important to be aware of the variables which are present in this admittedly 'artificial' system for cell growth, and the possible importance of such variables in the study of genetic disorders. It will be the purpose of this review to discuss the importance of cell culture variables within the scope of the study of metabolic diseases. It will be necessary to present a brief discussion of cell culture media in general, before going on to review studies with particular relevance to our topic.

GENERAL CONSIDERATIONS

The environment of cultured human fibroblasts is the culture medium, the substrate or support on which the cells grow, neighbouring cells and the gas phase. I shall limit my discussion to the culture medium. Several reviews have discussed the development of tissue culture media which are in common use today (Paul, 1965; Paul, 1975; Waymouth, 1965). In the beginning of the tissue culture era, cells were grown in a milieu as close as possible to that of the body

* Supported by Grants No C.80.53–2 and C.30.53–12 from The Norwegian Research Council for Science and the Humanities

fluids. Gradually, synthetic formulas were added, either as supplements to body fluids, or later as the main component to which body fluids were added as the supplement. Finally, completely synthetic media were prepared which could promote growth of at least some established cell strains (Waymouth, 1965).

The basic components of a synthetic medium are a salt mixture, an energy source, amino acids and some vitamins. In addition, the pH must be maintained within an acceptable range, and serum of some sort is necessary for diploid cell growth and division. An example of such a medium is Eagle's minimal essential medium, the components of which are listed in Table 2.1.

Salt mixture

The cells must be maintained in an environment of appropriate osmotic pressure (usually provided by sodium chloride). However, quite severe changes in osmotic pressure ($\pm 10\%$) are tolerated by human fibroblasts without marked effect on cell metabolism (Paul, 1975). The other ions provided in the culture medium are necessary components in cell metabolism. Inorganic phosphate seems to be rate-limiting for glycolysis if organic phosphates such as those present in biological fluids are not present in the medium (Waymouth, 1954). Mg^{2+} and Ca^{2+} are required to anchor the cells to the growth surfaces, and have important functions in intermediary metabolism. Mg^{2+}, for instance, is directly involved in all regulated metabolic pathways, since it is necessary for all transphosphorylation reactions. A central role of Mg^{2+} in the control of cell growth has been proposed by Rubin (1975) and Rubin and Koide (1976).

Trace metals such as Zn^{2+} have also been thought necessary for normal cell growth (Rubin, 1973). Iron is required for the synthesis of respiratory enzymes. The need for other trace elements is at present unclear. However, a genetic deficiency of copper metabolism has recently been detected in cells taken from patients with Menke's syndrome (Goka et al., 1976), indicating that the cells recognize this element in a genetically controlled fashion.

Energy source

Cultured cells need energy for metabolic processes and mechanical work such as locomotion (Paul, 1965). The widely used energy source is glucose, although other mono- or disaccharides (e.g. fructose or mannose) can replace it. The metabolism of glucose is mainly glycolytic, resulting in the accumulation of lactic acid in the medium. However, cells certainly do possess all the enzyme machinery needed for respiration and probably also for the pentose shunt (Paul, 1965). The balance between glycolysis and respiration can be varied according to cultural conditions. For instance, inhibition of respiration by anaerobic conditions causes increased glycolysis (Pasteur effect), while increased concentrations of glucose may inhibit respiration (Crabtree effect). pH changes will also influence glucose metabolism, since it has been known for a long time that

Table 2.1   Minimal essential medium of Eagle (Eagle and Lavintow, 1965)

Compound	Concentration (mM)	(mg/1)	Compound	Concentration (mM)	(mg/1)
L-amino acids			**Salts**		
Arginine	0.6	105	NaCl	116	6800
Cystine	0.1	24	KCl	5·4	400
Glutamine	2.0	292	CaCl$_2$	1·8 (0)*	200 (0)*
Histidine	0.2	31	MgCl$_2$·6H$_2$O	1·0	200
Isoleucine	0.4	52	NaH$_2$PO$_4$·2H$_2$O	1·1 (11)*	150 (1500)
Leucine	0.4	52	NaHCO$_3$	23·8	2000*
Lysine	0.4	58			
Methionine	0.1	15			
Phenylalanine	0.2	32	**Vitamins**		
Threonine	0.4	48	Choline		1
Tryptophan	0.05	10	Folic acid		1
Tyrosine	0.2	36	Inositol		2
Valine	0.4	46	Nicotinamide		1
			Pantothenate		1
Carbohydrate			Pyridoxal		1
Glucose	5.5	1000	Riboflavin		0.1
			Thiamine		1
Serum protein					
Whole or dialyzed serum, 5–10%					

* For suspension cultures.

cells grown at an alkaline pH produce much more lactic acid than cells grown at a low pH (Zwartouw and Westwood, 1958). This has been found in many cell types and is not necessarily connected with a corresponding change in respiration (Paul, 1965). The external glucose concentration will also markedly influence the pattern of glucose metabolism, since a progressive increase in glucose concentration promotes glycolysis (Paul, 1965).

The main regulatory enzyme in glycolysis is phosphofructokinase. It has been reported that the activity of this enzyme increases markedly when resting cells are stimulated to grow (Fodge and Rubin, 1973). Aerobic glycolysis in lymphocytes has recently been shown to be closely linked to periods of active DNA synthesis (Wang et al., 1976). If a shift towards alkaline pH is considered as a growth stimulus, this might be the explanation of the pH dependent lactate production referred to above.

Amino acids

No cultured mammalian cell has yet been shown to grow in the absence of the eight essential amino acids usually required by man (Eagle and Lavintow, 1965). This means that the biosynthetic machinery of these amino acids missing in the body is certainly not acquired during the adaptation to *in vitro* growth. However, five additional amino acids are also required for cellular growth *in vitro*. The requirements of glutamine, arginine and tyrosine are not surprising, since these are mainly synthesized by the liver in the intact organism. The same may be true for histidine, which is an essential amino acid in infants (Snyderman et al., 1957). The requirement of cyst(e)ine is more difficult to explain. Non-essential amino acids are often added to the minimal medium when more delicate cells or sparse cultures require to be established.

Vitamins

Eight vitamins have been demonstrated to be required by cultured cells, and are commonly included in media, although some media contain more. Six of these vitamins have functions as coenzymes in important steps in cell metabolism and their indispensibility is hardly surprising (Eagle and Lawintow, 1965). The role of specific vitamins in the studies of genetic disease will be discussed later.

pH and CO_2 concentration

Cells in culture must be maintained within small pH limits in order to function properly. The buffering system most widely used until recently has been carbonic acid/bicarbonate in equilibrium with 5% CO_2 in the gas phase. Because of difficulties with such an artificial gas phase, effort has been devoted to the development of buffers not dependent upon CO_2. Recently, much work has been done with the organic buffers originally described by Good *et al.* (1966) and applied in tissue culture by Eagle (1971). However, it seems quite clear that during prolonged cell growth CO_2 itself is essential (Paul, 1975). Cells can grow in organic buffers for a while, but after prolonged culture there is reduced growth. One suggested explanation of this phenomenon has been that the absence of CO_2 causes decarboxylation of oxaloacetate, with a consequent cessation of cell respiration. In support of this theory it has been reported that added oxaloacetate may restore growth in CO_2 deprived cultures (Gwatkin and Siminovitch, 1960).

The effect of pH changes on glycolysis has already been mentioned. Other effects of pH on metabolic characteristics relevant to the study of metabolic disease will be discussed later.

Serum

Although complete synthetic media can be used for the propagation of certain heteroploid established cell strains, we certainly still need serum for growth of human diploid fibroblasts. The growth promoting effect of serum has been discussed earlier in this symposium, and just a few recent observations on the effect of serum on cell metabolism will be mentioned. In confluent cultures, one of the earliest observable effects is an increased transport rate of various nutrients (Pardee, 1975). In chick embryo fibroblasts, there is an immediate increase in the rate of glucose transport, followed hours later by an increase in protein and DNA synthesis (Sefton and Rubin, 1971). Such increases are seen even when protein as well as RNA synthesis is inhibited (Rubin and Koide, 1975). In 3T3 cells, uridine and phosphate transport are increased several fold within minutes, followed by subsequent DNA synthesis and cell division. This has been shown to be due to a change in the V_{max} but not in the K_m of the transport machinery (Stein and Rosengurt, 1976). Cyclic AMP or cGMP levels may be involved in the growth promoting effect of serum, since it has been shown that the activity of cyclic nucleotide phosphodiesterase increases rapidly after serum has been added to resting cells (Pledger *et al.*, 1975).

It is important to have in mind these and possibly other serum effects when studying cell metabolism, since the frequency of medium change or the concentration of serum used (Felix *et al.*, 1974) may affect the intracellular pool of various nutrients and their metabolism. One specific effect of serum on cell behaviour is that reported by Kittlich *et al.* (1973). They could show that the pattern of mucopolysaccharide synthesis in rat embryonic fibroblasts was influenced by the concentrations of serum used.

CELL CULTURE MEDIUM AND THE INBORN ERRORS OF METABOLISM

Today we are faced with a wide variety of commercially available media using different types and concentrations of biological fluids as supplements. We now turn to a discussion of variables which might be of specific importance in the study of genetic disease, whether studying specific enzyme activities or metabolic functions in intact cells. It is outside my task to discuss variables intrinsic to the cells as such, like stage of growth, cell densities or passage number, although these variables clearly are very relevant to the topic under discussion. I will follow the list of variables already given and emphasize only those related to genetic disease.

Salt mixture

Variation in medium Mg^{2+} concentration has been reported to affect the activity of hypoxanthine guanine phosphoribosyltransferase (Benke *et al.*, 1973), the mutant enzyme in Lesch–Nyhan syndrome. Addition of small, but non-toxic amounts of magnesium increased the enzymatic activity in mutant cells. The probable explanation offered was that the concentration of Mg^{2+} in the commercially obtained F-12 medium was 6×10^{-4} M, which is only one-third of the binding constant of the Mg-phosphoribosyl-1-pyrophosphate to the enzyme. Other enzymes may well show similar dependence on divalent cation concentrations.

Energy source

Although glucose is the main energy source in most culture media, other sugars have been used in specific studies. The first metabolic deficiency detected in fibroblasts was galactosaemia. Krooth and Weinberg showed in 1960 that cells from galactosaemic patients were unable to use galactose as the sole energy source. Relevant to this topic, however, is the demonstration that galactose in the growth medium will stimulate galactokinase activities in fibroblast cultures both from normal individuals and heterozygotes for this gene (Zacchello *et al.*, 1972). Homozygous affected individuals, did not show any increased activity when galactose was added to the medium. This observation might indicate that induction of enzyme activities is possible in other systems. Clearly, such an induction could minimize differences between homozygous normal and heterozygote individuals.

Amino acids

I am not aware of any studies which have shown that differences in the amino acid composition of culture medium has affected metabolic processes relevant to the study of genetic diseases.

Vitamins

The vitamin content of various synthetic media varies a great deal and this can influence the results of studies in metabolic disease.

Vitamin C is known to be indispensable for production of collagen by human fibroblasts in tissue culture. However, vitamin C has also been reported to affect intracellular content of mucopolysaccharides differently in normal cells as compared to cells taken from patients with Hurler's syndrome (Schafer *et al.*, 1966). Vitamin C caused a marked increase in intracellular mucopoly-saccharides in the mutant cell but a less marked change in the normal cells. Today we can explain this observation if we assume that the presence of vitamin C increases cellular synthesis of mucopolysaccharides.

Willcox and Patrick have developed a very sensitive and rapid diagnostic method for cystinosis using the incorporation of L-[^{35}S]cystine into fibroblasts (Willcox and Patrick, 1974). Mutant cells incorporate more cystine than normal controls. However, Kroll and Schneider (1974) have shown that addition of L-ascorbic acid to the culture medium decreases the content of free cystine by more than 50% in cystinotic fibroblasts.

The B-vitamin responsive inborn errors of metabolism are well known, and it is possible that this responsiveness is reflected in tissue culture. Nadler and his group have demonstrated this to be the case in methylmalonic acidaemia, where the addition of 5'-deoxyadenosylcobalamin corrected the metabolic deficiency in cells taken from a patient with this disorder (Kaye *et al.*, 1974). However, in this particular study, the correction *in vitro* was not seen *in vivo*, since the patient failed to respond to high doses of vitamin B_{12}. Certainly, other effects of B vitamins on enzyme activities in vitamin B-responsive metabolic deficiencies may well appear.

Hydrogen ion concentrates

The pH of the cell culture medium is perhaps the variable which is most difficult to control. In recent years, much work has been devoted to studies on possible pH dependence of various cell characteristics. Table 2.2 lists some properties of cultured cells which have shown marked dependence on the pH of the medium.

Effect on cell morphology

One of the first reported pH effects came long before the use of organic buffers, and was Taylor's study on cell morphology and locomotion (Taylor, 1962). We have confirmed these findings, and have demonstrated marked pH effects on cell morphology. Cells at an alkaline pH tend to be somewhat larger and more spherical, while cells in a more acid milieu tend to flatten out. This is well demonstrated using dark-field microscopy (Figure 2.1a and b). Cells at an alkaline pH tend to have a smaller contact with the growth surface. Perhaps

Table 2.2 Properties of cultured cells dependent on the pH of the medium

Property	Reference
Cell morphology and locomotion	Taylor, 1962
Glycolysis	Zwartouw and Westwood, 1958
Cell growth	Ceccarini, 1975; Ceccarini and Eagle, 1971a, 1971b; Eagle, 1973; Froehlich and Anastassiades, 1974; Lie et al., 1972; Rubin, 1971, 1973
Membrane transport	Ceccarini and Eagle, 1971a; Kroll and Schneider, 1974; Rubin and Koide, 1975
Mucopolysaccharide metabolism	Kittlick and Neupert, 1972; Lie, 1974; Lie et al., 1972, 1973; Schwartz et al., 1976
Collagen production	Nigra et al., 1973
Virus yield	Calothy et al., 1973; Fields and Eagle, 1973; Gierthy et al., 1974
Cell hybridization	Croce et al., 1972

because of this they are much more easy to trypsinize than cells grown at a more acid pH. It could also well be that a relatively larger part of the cell membrane is in contact with the alkaline medium, thus contributing to increased transport rates.

Effects on cell growth

It has been known for a long time that the pH of the medium has a marked effect on cell metabolism—for instance high pH promotes lactate production (Zwartouw and Westwood, 1958). In 1971, Eagle and coworkers introduced organic buffers in order to improve control of pH and to study its effect on cell behaviour (Eagle, 1971). Ceccarini and Eagle showed that a rise in pH stimulated uridine transport and subsequent cell growth (Ceccarini and Eagle, 1971). Contact-inhibited cells could be released from their inhibition simply by making the medium more alkaline, and the cells reached much higher final densities. Most normal cells of human origin have growth optima between pH 7.6–7.8, while malignant cells in general are less sensitive to pH changes (Eagle, 1973). The growth-promoting effect of alkaline pH is probably more pronounced in confluent than in sparse cultures (Froehlich and Anastassiades, 1974).

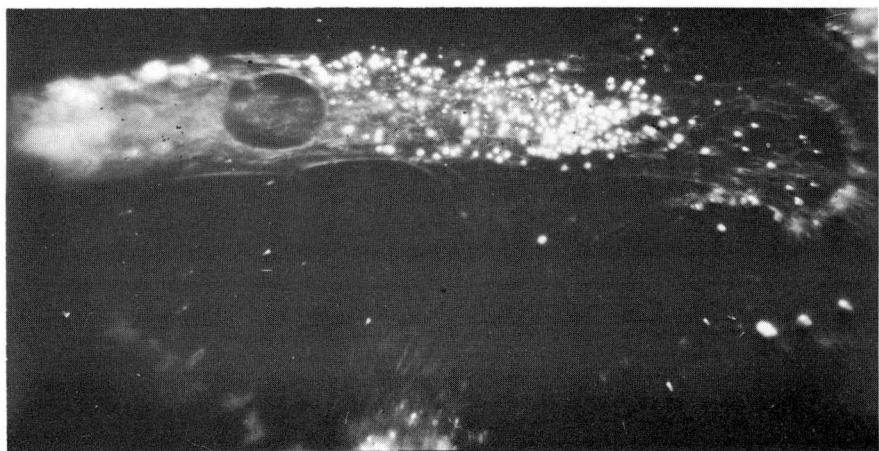

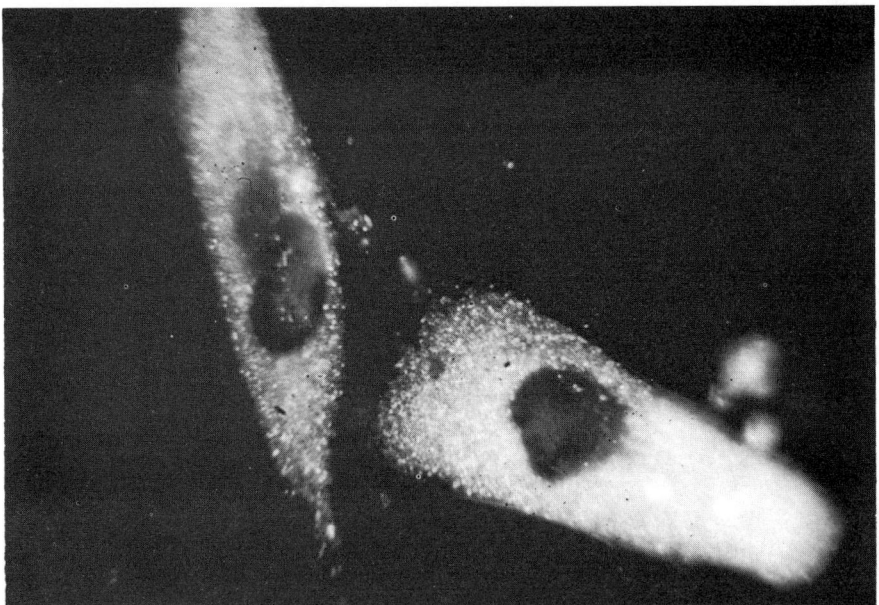

Figure 2.1   Dark-field microscopy of human fibroblasts in culture grown at pH 6.8 (Figure 1a) and pH 7.8 (Figure 1b) for 3 days

Effect on membrane transport

Transport of nutrients into the cells is pH dependent. Alkaline pH increases the influx of various nutrients in different cell types (Ceccarini and Eagle, 1971; Rubin and Koide, 1975). In a discussion on lysosomotropic agents, de Duve *et al.* (1974) proposed that this is probably true for all weak bases, although in this connection the increased transport rates are caused by changes in the proportion of uncharged molecules able to penetrate lipid bilayers rather than to changes in the cells themselves. However, uptake of macromolecules through the process of endocytosis might also show pH dependence. We have shown that uptake of adriamycin complexed with DNA is strongly pH dependent (Lie and Lie, 1977), but have no evidence as yet that this is the case for other macromolecules which presumably enter cells by endocytosis. (See Figure 2.2.)

Effect on mucopolysaccharide (MPS) metabolism

I shall describe in some detail how one can mimic a storage disease by manipulating medium pH. Neufeld and her coworkers defined the genetic deficiencies in MPS degradation using labelled inorganic sulphate in the medium in order to follow the MPS metabolism in cultured cells (see Neufeld and Cantz, 1973 for review). This synthesis is very efficient in human fibroblasts, and kinetic analyses have shown that approximately 75% of the newly synthesized molecules are secreted from the cell, while 25% remain inside, mainly in the lysosomal fraction. The latter must be broken down before the label can escape, and the method is a very sensitive indicator of storage of MPS inside the cells. Neufeld and co-workers could show that for all the genetic deficiencies in MPS degradation (except type 4) there is a progressive increase in intracellular label, while normal cells reach a steady state after some hours. They could furthermore show cross-correction between different genotypes, e.g. a normal metabolic pattern of MPS was obtained when two different mutant cell lines were cocultivated. Corrective factors could be isolated for each of the mucopolysaccharidoses which could at that time be studied at the cellular level. Today, of course, the basic defect in these disorders are well characterized (Dorfman and Matalon, 1976; Neufeld et al., 1975).

Several laboratories have had difficulties in reproducing the elegant work of Neufeld and co-workers. We found that the probable reason for these difficulties was that the intracellular MPS degradation is highly pH dependent, becoming progressively inhibited as medium pH increases (Lie et al., 1972). Indeed, at pH 7.8–7.9, all normal human fibroblasts behave in this system as if they have a genetic deficiency in MPS degradation. Synthesis and release of MPS to the surrounding media was not influenced by medium pH in contrast to the findings reported for embryonic rat fibroblasts (Dean et al., 1976). This would indicate that the pH effect on lysosomal function was rather specific. Degradation of MPS in pH inhibited cells started immediately after returning the pH to a more acid range, and this reactivation was unaffected by inhibitors of protein synthesis, indicating that the enzymes necessary for MPS degradation were present, but in an inactive form. Recently, results similar to our own have been obtained in chondrocyte cultures (Schwartz et al., 1976). The pH effect in this system was so pronounced that the authors speculated about possible in vivo importance of this sensitivity.

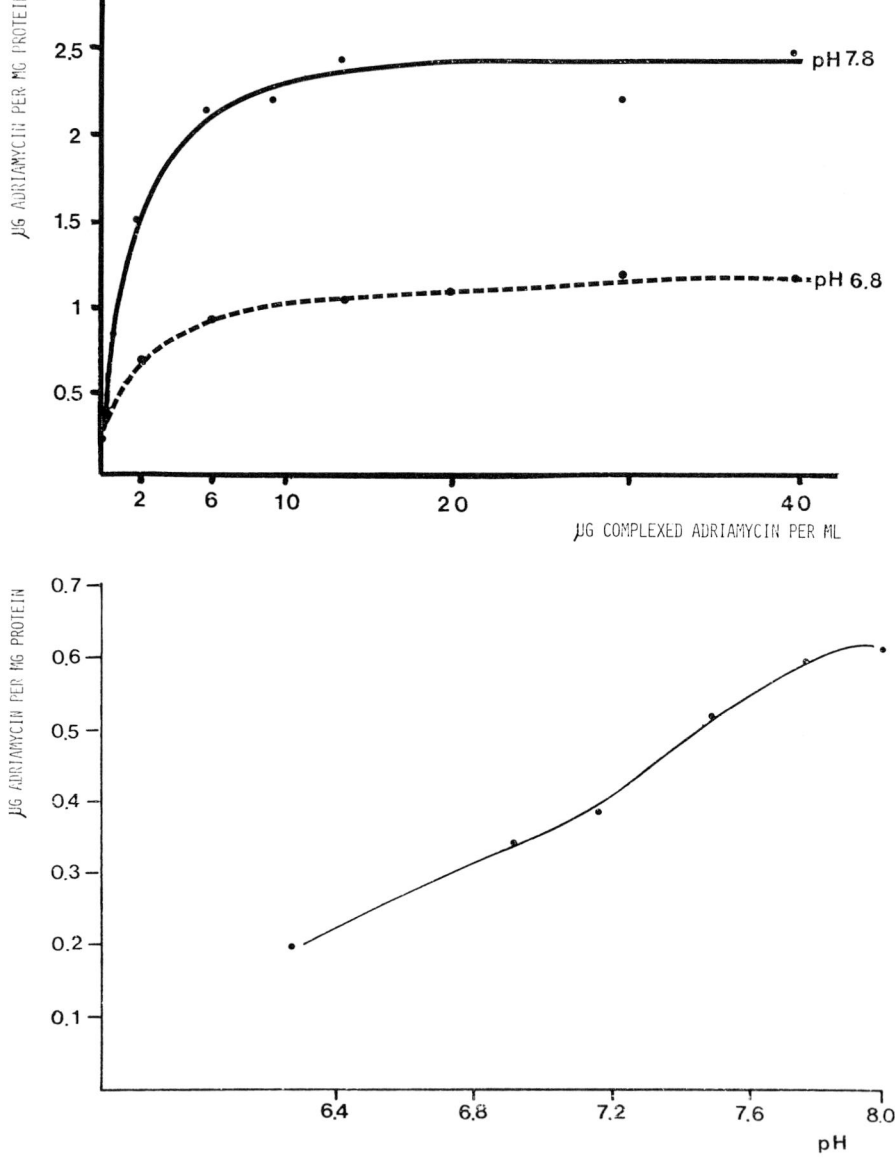

Figure 2.2   pH dependence of uptake of DNA-complexed adriamycin by human fibroblasts in culture. Prior to uptake, the cells were grown at identical pHs and the uptake studies performed at the pH indicated

We have no obvious explanation of this pH effect. One possibility is that it might be due to an increase in intralysosomal pH to the point where pH inactivation of crucial enzyme activities could occur. We have shown that several weak bases which tend to concentrate in cell lysosomes (chloroquine, quinine, acridine orange and neutral red) will inactivate this function of MPS degradation (Lie, 1974; Lie and Schofield, 1973). Although these substances may act by increasing intralysosomal pH, they reach concentrations inside the lysosomes which might in themselves be inhibitory to enzyme activities (de Duve et al., 1974).

Certainly, these artificially produced 'storage' cells show an abundance of lysosome-like structures on electron microscopy, some of which are acid phosphatase positive (Lie and Schofield, 1973; Lie et al., 1973). Indeed, the acid phosphatase reaction in the cells grown at alkaline pH was much stronger than corresponding activity in the same genotype cultured at a lower pH. This paradoxical situation illustrates that demonstration of enzyme activities in cell homogenates or the presence of intracellular enzyme by histochemical techniques does not necessarily mean that this activity is present in the living cell.

Effect on enzyme activities in cell homogenates

Ryan et al. found that β-glucosidase and aryl sulphatase activities were twice as high in cultures grown at pH 7.3 compared with pH 6.8 (Ryan et al., 1972). Wood (1975), studying the effect of environmental pH upon acid hydrolase activities in human diploid fibroblasts observed significantly decreased activities of α-galactosidase, β-galactosidase, α-glucosidase, β-glucuronidase and hexosaminidase with increasing pH, while β-glucosidase increased with the same pH changes. Such pH-dependent variations were not seen by Butterworth et al. (1974) who used cultured amniotic fluid cells. This difference may be due to the different cell types studied. However, I would anticipate that pH effects on metabolic characteristics are more pronounced in intact cells than in cell homogenates.

Serum

In HeLa cells, lysosomal acid phosphatase, but not β-glucuronidase or N-acetyl-β-D-hexosaminidase increases with increasing serum concentrations (Wang and Touster, 1976), although other lysosomal activities are not influenced by the various growth stages of these cells as measured in synchronized cultures (Berg et al., 1975). Although serum concentrations were obviously correlated with cell growth, Butterworth and co-workers did not observe any effect of serum on specific activities of ten lysosomal enzymes in cultured amniotic fluid cells, although β-galactosidase tended to be lower at high (30%) concentrations of fetal calf serum (Butterworth et al., 1974).

Variations in lysosomal enzyme activities which are known to occur in cultured cells are very intriguing. For instance, Milunsky and co-workers have found that wide fluctuations can be observed in the same human fibroblasts studied in parallel under identical cultural conditions (Milunsky *et al.*, 1972). For each of five lysosomal enzymes studied, there was a 50–600% variation in the enzymatic activities, and in one analysis a level of aryl sulphatase was so low as to suggest the diagnosis of homozygous metachromatic leukodystrophy. Faced with such wide and unpredictable variations in enzyme activities, it has been proposed that enzyme ratios be used rather than specific activities in the diagnosis of lysosomal disorders (Lindsten *et al.*, 1976; Young *et al.*, 1975).

The serum-dependence of non-lysosomal enzymes are not well studied. However, Mellman *et al.* have shown that the turnover of catalase is dependent upon a serum factor (Mellman *et al.*, 1972).

CORRECTION OF METABOLIC DEFICIENCIES IN CULTURED CELLS BY CHANGES IN CULTURE MEDIA

One important question that has to be considered by all investigators studying inborn errors of metabolism in cell culture, is whether cultural conditions might artificially correct a defect. This is well-documented in the studies of Neufeld on corrective factors and mucopolysaccharidoses (Neufeld and Cantz, 1973; Neufeld *et al.*, 1975). Conditioned media from normal or another mutant cell have enough corrective activity to normalize the abnormal MPS storage in mutant cells, and this is due to absorptive pinocytosis of the missing enzyme by the deficient cell. Conditioned media have also been reported to reduce the storage of lipid observed in cells taken from patients with Wolman's disease (Kyriakides, 1972). Purified enzymes will correct the metabolic defect in cells from patients with Fabry's disease (Dawson *et al.*, 1973) but not in cells from patients with Tay–Sach's disease (Schneck *et al.*, 1973). It has been recommended that human serum should not be used in the culture of cells used in genetic diagnosis because of the possible presence of corrective factors.

Elegant work by Reuser and co-workers using enzyme assays in single human fibroblasts after cocultivation of normal and deficient cells has demonstrated correction of deficient cells. Using various mutant cells in mixed culture with normal cells, they could show transfer of N-acetyl-β-D-hexosaminidase, but not of acid α-glucosidase and β-galactosidase. These studies most closely reflect the *in vivo* situation and are of importance in the therapy of the lysosomal disorders by enzyme replacement (Desnick *et al.*, 1976; Herd *et al.*, 1976; Tager *et al.*, 1974). This is highlighted by the recent observations of Dean and co-workers who could show a beneficial effect from subcutaneous transplantation of normal cultured fibroblasts in a patient with the Hunter syndrome (Dean *et al.*, 1975; 1976).

CONTAMINATION

One final note of caution. There is one way in which metabolic deficiencies can be 'corrected' in cultured cells, and that is by mycoplasma infection. This nuisance found in all laboratories working with cells in culture has been known for some time to cause altered metabolism of nucleic acids (Levine *et al.*, 1968) and to cause chromosomal abnormalities (Schneider *et al.*, 1974). However, recently it has been shown that the presence of mycoplasma species in cultured cells can produce normal enzymic activities in deficient cells. Stanbridge *et al.* (1975) found almost normal levels of hypoxanthine guanine phosphoribosyltransferase when cells deficient in this enzyme were contaminated with mycoplasma. Increased activities of hexosaminidase A were also found. Clark and co-workers have recently reported that pyruvate dehydrogenase activity is high in mycoplasma, and could show that that a low degree of contamination increased the activity of this enzyme by 1000% in deficient cells (Clark *et al.*, 1976). Mycoplasma should therefore always be considered as a potential source of enzyme and continuous scrutiny of cell cultures for the organism is therefore important (Harnett *et al.*, 1974; Russell *et al.*, 1975).

CONCLUSIONS

This paper has discussed the importance of variations in the cell culture medium on fibroblast characteristics relevant to the study of genetic disease. The well known clinical problem of establishing the relative importance of heredity and environmental factors in an individual's phenotype may also apply to cells in culture. The evidence presented emphasizes the generally accepted conclusion that biochemical diagnosis of genetic deficiencies in tissue culture needs to be studied under strictly controlled conditions and with proper controls matched for age, cell densities, passage number and culture conditions.

References

Benke, P. J., Hebert, A. and Herrick, N. (1973). *In vitro* effects of magnesium ions on mutant cells from patients with the Lesch–Nyhan syndrome. *N. Engl. J. Med.*, **289**, 446

Berg, T., Melbye, B., Johnsen, S. R. and Prydz, H. (1975). Activity of lysosomal enzymes in the various cell cycle phases of synchronized HeLa cells. *Exp. Cell Res.*, **94**, 106

Beutler, E., Kuhl, W., Trinidad, F., Teplitz, R. and Nadler, H. (1971). Beta-glucosidase activity in fibroblasts from homozygotes and heterozygotes for Gaucher's disease. *Am. J. Hum. Genet.*, **23**, 62

Butterworth, J., Sutherland, G. R., Broadhead, D. M. and Bain, A. D. (1974). Effect of serum concentration, type of culture medium and pH on the lysosomal enzyme activity of cultured human amniotic fluid cells. *Clinica Chimica Acta*, **53**, 239

Calothy, G., Groce, C. M., Defendi, V., Koprowski, H. and Eagle, H. (1973). Effect of environmental pH on rescue of Simian virus 40. *Proc. Nat. Acad. Sci. of the U.S.A.*, **70**, 366

Ceccarini, C. (1975). Effect of pH on plating efficiency, serum requirement, and incorporation of radioactive precursors into human cells. *In vitro*, **11**, 78

Ceccarini, C. and Eagle, H. (1971a). Induction and reversal of contact inhibition of growth by pH modification. *Nature (London) New Biol.*, **233**, 271

Ceccarini, C. and Eagle, H. (1971b). pH as a determinant of cellular growth and contact inhibition. *Proc. Nat. Acad. Sci. U.S.A.*, **68**, 229

Clark, A. F., Farrell, D. F. and Scott, C. R. (1976). The effect of mycoplasma contamination on the *in vitro* assay of pyruvate dehydrogenase activity on cultured fibroblasts. *Clin. Res.* **24,** 147A (Abstr.)

Groce, C. M., Koprowski, H. and Eagle, H. (1972). Effect of environmental pH on the efficiency of cellular hybridization. *Proc. Nat. Acad. Sci. U.S.A.,* **69,** 1953

Cunningham, D. D. and Pardee, A. B. (1969). Transport changes rapidly initiated by serum addition to 'contact inhibited' 3T3 cells. *Proc. Nat. Acad. Sci. U.S.A.,* **64,** 1049

Dawson, G., Matalon, R. and Li, Y. T. (1973). Correction of the enzymic defect in cultured fibroblasts from patients with Fabry's disease: treatment with purified alpha-galactosidase from ficin. *Pediat. Res.,* **7,** 684

Dean, M. F., Muir, H., Benson, P. F., Button, L. R., Batchelor, J. R. and Bewick, M. (1975). Increased breakdown of glycosaminoglycans and appearance of corrective enzyme after skin transplants in Hunter syndrome. *Nature (London),* **257,** 609

Dean, M. F., Muir, H., Benson, P. F., Button, L. R., Boylston, A. and Mowbray, J. (1976). Enzyme replacement therapy by fibroblast transplantation in a case of Hunter syndrome. *Nature (London),* **261,** 323

de Duve, C., De Barsy, T., Poole, B., Trouet, A., Tulkens, P. and Van Hoof, F. (1974). Lysosomotropic agents. *Biochem. Pharmacol.,* **23,** 2495

Desnick, R. J., Thorpe, S. R. and Fiddler, M. B. (1976). Toward enzyme therapy for lysosomal storage diseases. *Physiol. Rev.,* **56,** 57

Dorfman, A. and Matalon, R. (1976). The mucopolysaccharidoses. *Proc. Nat. Acad. Sci. U.S.A.,* **73,** 630

Eagle, H. (1971). Buffer combinations for mammalian cell culture. *Science,* **174,** 500

Eagle, H. (1973). The effect of environmental pH on the growth of normal and malignant cells. *J. Cell. Physiol.,* **82,** 1

Eagle, H. and Lavintow, L. (1965). Amino acid and protein metabolism. In: *Cells and Tissues in Culture,* Vol. 1, pp. 277–296. E. N. Willmer (ed). (London, New York: Academic Press)

Felix, J. S., Doherty, R. A. Davis, H. T. and Tidge, S. C. (1974). Amniotic fluid cell culture. 1. Experimental design for evaluating cell culture variables: determination of optimal fetal calf serum concentration. *Paediat. Res.,* **8,** 870

Fields, B. N. and Eagle, H. (1973). The pH-dependence of reovirus synthesis. *Virology,* **52,** 581

Fodge, D. W. and Rubin, H. (1973). Activation of phosphofructokinase by stimulants of cell multiplication. *Nature (London) New Biol.,* **246,** 181

Froehlich, J. E. and Anastassiades, T. P. (1974). The role of pH in fibroblast proliferation. *J. Cell. Physiol.,* **84,** 253

Gierthy, J. F., Ellem, K. A. O. and Singer, I. I. (1974). Environmental pH and the recovery of H-1 parvovirus during single cycle infection. *Virology,* **60,** 548

Goka, T. J., Stevenson, R. E., Hefferan, P. M. and Howell, R. R. (1976). Menke's disease: a biochemical abnormality in cultured human fibroblasts. *Proc. Nat. Acad. Sci. U.S.A.,* **73,** 604

Good, N. E., Winget, G. D., Winter, W., Connolly, T. N., Izawa, S. and Singh, R. M. M. (1966). Hydrogen ion buffers for biological research. *Biochemistry,* **5,** 467

Gwatkin, R. B. L. and Siminovitch, L. (1960). Multiplication of single mammalian cells in a non-bicarbonate medium. *Proc. Soc. Exp. Biol. Med.,* **103,** 718

Harnett, G. B., Phillips, P. A. and Mackay-Scollay, E. M. (1974). A simple method for detecting mycoplasma infection of cell cultures. *Clin. Pathol.,* **27,** 70

Herd, J. K., Hayhome, B. A., Tschida, J. and Forrest, T. (1976). *In vitro* correction of Hurler fibroblasts with bovine testicular hyaluronidase. *Proc. Soc. Exp. Biol. and Med.,* **151,** 642

Kaye, C. I., Morrow, G., III and Hadler, H. L. (1974). *In vitro* 'responsive' methylmalonic acidemia: a new variant. *J. Pediat.,* **85,** 55

Kittlick, P.-D. and Neupert, G. (1972). Experimentelle Beeinflussing der Synthese saurer Mucopolysaccharide (Glykosaminoglykane) in Fibroblasten kulturen. IV Einfluss des pH wertes auf Zellproliferation und MPS—muster. *Experimentelle Pathologie,* **7,** 117

Kittlick, P.-D., Neupert, G. and Lumkemann, U. (1973). Die Wirkung reischiedener seren auf die Mucopolysaccharidsynthese in Fibroblasten kulturen. *Experimentelle Pathologie*, **8**, 194

Kroll, W. A. and Schneider, J. A. (1974). Decrease in free cystine content of cultured cystinotic fibroblasts by ascorbic acid. *Science*, **186**, 1040

Krooth, R. S. and Weinberg, A. M. (1960). Properties of galactosemic cells in culture. *Biochem. Biophys. Res. Commun.*, **3**, 518

Kyriakides, E. C., Paul, B. and Balint, J. A. (1972). Lipid accumulation and acid lipase deficiency in fibroblasts from a family with Wolman's disease, and their apparent correction *in vitro*. *J. Lab. Clin. Med.*, **80**, 810

Leroy, J. G. and DeMars, R. I. (1967). Mutant enzymatic and cytological phenotypes in cultured human fibroblasts. *Science*, **157**, 804

Levine, E. M., Thomas, L., McGregor, D., Hayflick, L. and Eagle, H. (1968). Altered nucleic acid metabolism in human cell cultures infected with mycoplasma. *Proc. Nat. Acad. Sci. U.S.A.*, **60**, 583

Lie, K. K. and Lie, S. O. (1976). Studies on cellular uptake of free and complexed adriamycin. In *Proceedings of the 9th Congress of the Nordic Society for Cell Biology*, pp. 121–124. F. Bierring (ed.) Odense University Press

Lie, S. O. (1974). Induction of 'storage disease' in normal human fibroblasts. In: *Birth Defects*; Original Article Series, **10**, 203

Lie, S. O., McKusick, V. A. and Neufeld, E. F. (1972). Simulation of genetic mucopolysaccharidoses in normal human fibroblasts by alteration of pH of the medium. *Proc. Nat. Acad. Sci. U.S.A.*, **69**, 2361

Lie, S. O. and Schofield, B. (1973). Inactivation of lysosomal function in normal cultured human fibroblasts by chloroquine. *Biochem. Pharmacol.*, **22**, 3109

Lie, S. O., Shofield, B. H., Taylor, H. A., Jr. and Doty, S. B. (1973). Structure and function of the lysosomes of human fibroblasts in culture: dependence on medium pH. *Paediat. Res.*, **7**, 13

Lindstèn, J., Zetterström, R. and Ferguson-Smith M. (eds.) (1976). Prenatal diagnosis of genetic disorders of the foetus. *Acta Paediat. Scand.*, Suppl. 259

Mellman, W. J., Schimke, R. T. and Hayflick, L. (1972). Catalase turnover in human diploid cell cultures. *Exp. Cell Res.*, **73**, 399

Milunsky, A., Spielvogel, C. and Kanfer, J. N. (1972). Lysosomal enzyme variations in cultured normal skin fibroblasts. *Life Sci.*, **11**, 1101

Nuefeld, E. F. and Cantz, M. (1973). In: *Lysosomes and Storage Diseases*, H. G. Hers and I. van Hoof (eds.), p. 262 (New York and London: Academic Press)

Neufeld, E. F., Lim, T. W. and Shapiro, L. J. (1975). Inherited disorders of lysosomal metabolism. *Annu. Rev. Biochem.*, **44**, 357

Nigra, T. P., Martin, G. R. and Eagle, H. (1973). The effect of environmental pH on collagen synthesis by cultured cells. *Biochem. Biophys. Res. Commun.*, **53**, 272

Pardee, A. B. (1975). The cell surface and fibroblast proliferation some current research trends. *Biochim. Biophys. Acta*, **417**, 153

Paul, J. (1965). Carbohydrate and energy metabolism. In: *Cells and Tissues in Culture*, Vol. 1, pp. 239–276. E. N. Willmer (ed.) (London and New York: Academic Press)

Paul, J. (1975). *Cell and Tissue Culture*. 5th Ed. (London and Edinburgh: Churchill Livingstone)

Pledger, W. J., Thompson, W. J. and Strada, S. J. (1975). Mediation of serum-induced changes in cyclic nucleotide levels in cultured fibroblasts by cyclic nucleotide phosphodiesterase. *Nature (London)*, **256**, 729

Reuser, A., Halley, D., De Wit, E., Hoogeveen, A., van der Kamp, M., Mulder, B. and Galjaard, H. (1976). Intercellular exchange of lysosomal enzymes: enzyme assays in single human fibroblasts after co-cultivation. *Biochem. Biophy. Res. Commun.*, **69**, 311

Rubin, H. (1971). pH and population density in the regulation of animal cell multiplication. *J. Cell Biol.*, **51**, 686

Rubin, H. (1973). pH serum and Zn^{++} in the regulation of DNA synthesis in cultures of chick embryo cells. *J. Cell. Physiol.*, **82**, 231

Rubin, H. (1975). Central role for magnesium in co-ordinate control of metabolism and growth in animal cells. *Proc. Nat. Acad. Sci. U.S.A.*, **72**, 3551

Rubin, H. and Koide, T. (1975). Early cellular responses to diverse growth stimuli independent of protein and RNA synthesis. *J. Cell. Physiol.*, **86**, 47

Rubin, H. and Koide, T. (1976). Mutual potentiation by magnesium and calcium of growth in animal cells. *Proc. Nat. Acad. Sci. U.S.A.*, **73**, 168

Russell, W. C., Newman, C. and Williamson, D. H. (1975). A simple cytochemical technique for demonstration of DNA in cells infected with mycoplasmas and viruses. *Nature (London)*, **253**, 461

Ryan, C. A., Lee, A. Y. and Nadler, H. L. (1972). Effect of culture conditions on enzyme activities in cultivated human fibroblasts. *Exp. Cell Res.*, **71**, 388

Schafer, I. A., Sullivan, J. C., Svejcar, J., Kofoed, J. and Robertson, W. B. (1966). Vitamin C-induced increase of dermatan sulphate in cultured Hurler's fibroblasts. *Science*, **153**, 1008

Schneck, L., Amsterdam, D., Brooks, S. E., Rosenthal, A. L. and Volk, B. W. (1973). The Tay–Sach's disease fibroblast model: failure to respond to exogenous hexosaminidase. *Paediatrics*, **52**, 221

Schneider, E. L., Stanbridge, E. J., Epstein, C. J., Golbus, M., Abbo-Halbasch, G. and Rodgers, G. (1974). Mycoplasma contamination of cultured amniotic fluid cells: potential hazard to prenatal chromosomal diagnosis. *Science*, **184**, 477

Schwartz, E. R., Kirkpatrick, P. R. and Thompson, R. C. (1976). Effect of environmental pH on glycosaminoglycan metabolism by normal human chondrocytes. *J. Lab. Clin. Med.*, **87**, 198

Sefton, B. M. and Rubin, H. (1971). Stimulation of glucose transport in cultures of density-inhibited chick embryo cells. *Proc. Nat. Acad. Sci. U.S.A.*, **68**, 3154

Snyderman, S. E., Holt, L. E., Norton, P. M., Smellie, F. and Boyer, A. (1957). Valine and histidine: requirement of the normal infant. *Fed. Proc. Fed. Am. Soc. Exp. Biol.*, **16**, 399

Stanbridge, E. J., Tischfield, J. A. and Schneider, E. L. (1975). Appearance of hypoxanthine guanine phosphoribosyltransferase activity as a consequence of mycoplasma contamination. *Nature (London)*, **256**, 329

Stein, W. D. and Rosengurt, E. (1976). Temperature dependence of uridine transport in quiescent and serum-stimulated 3T3 cells. *Biochim. Biophy. Acta*, **419**, 112

Tager, T. M., Hooghwinkel, G. J. M. and Daims, W. T. (eds.) (1974). *Enzyme therapy in lysosomal storage diseases*. (Amsterdam: North-Holland Publishing Company)

Taylor, A. C. (1962). Responses of cells to pH changes in the medium. *J. Cell Biol.*, **15**, 201

Wang, C. C. and Touster, O. (1976). Acid phosphatase of HeLa cells: Properties and regulation of lysosomal activity by serum. *Arch. Biochem. Biophys.*, **172**, 191

Wang, T., Marquardt, C. and Foker, J. (1976). Aerobic glycolysis during lymphocyte proliferation. *Nature (London)*, **261**, 702

Waymouth, C. (1954). Some effects of inorganic phosphate and bicarbonate on cell survival and proliferation in chemically defined nutrient media. *Biochem. J.*, **56** (Proceedings abstr. iv–v)

Waymouth, C. (1965). Construction and use of synthetic media. In: E. N. Wilmer (ed.) *Cells and Tissues in Culture*, Vol. 1, pp. 99–142. (London and New York: Academic Press)

Willcox, P. and Patrick, A. D. (1974). Biochemical diagnosis of cystinosis using cultured cells. *Arch. Dis. Childhood*, **49**, 209

Wood, S. (1975). The effect of environmental pH upon acid hydrolase activities of cultured human diploid fibroblasts. *Exp. Cell Res.*, **96**, 317

Young, E., Willcox, P., Whitfield, A. E. and Patrick, A. D. (1975). Variability of acid hydrolase activities in cultured skin fibroblasts and amniotic fluid cells. *J. Med. Genet.*, **12**, 224

Zacchello, F., Benson, P. F., Brown, S., Croll, P., Gianelli, F. and Mann, T. P. (1972). Induction of galactokinase in fibroblasts from heterozygous and homozygous subjects. *Nature (London) New Biol.*, **239**, 95

Zwartouw, H. T. and Westwood, J. C. N. (1958). Factors affecting growth and glycolysis in tissue culture. *Br. J. Exp. Pathol.*, **39**, 529

3
Density-dependent Growth Regulation in Cultures of Human Diploid Fibroblasts

T. ELSDALE

Medical Research Council, Clinical and Population Cytogenetics Unit, Western General Hospital, Edinburgh

THE NEED FOR SERUM IN CELL CULTURE

It is well known that the survival and growth of fibroblasts in culture is dependent on the presence of serum in the medium. Recent observations show that the requirements for serum is greatly reduced using media containing trace elements and certain other metabolites in low concentrations in addition to the usual constituents (Mohamed *et al.*, 1976; Ham *et al.*, 1976). This suggests that serum may serve as a convenient source of recondite nutrients. It has, however, long been recognized that serum contains substances that are not cell nutrients in the strict sense, but are necessary in connection with substratum attachment, membrane stabilization and other aspects of cellular life *in vitro*. The current interest in investigations into the constituents of serum centres around proteinaceous substances of intermediate molecular weight (Temin *et al.*, 1972). The finding that some of these polypeptides or small proteins show hormone-like activities in appropriate assays suggests the possibility that they may affect the cell by inducing a configurational modification of the membrane at specific sites. This connects with another current interest in cell biology namely the functional heterogeneity of the cell surface.

Purified serum constituents customarily are assayed for three main activities: the potentiation of long term survival of cells, the stimulation of motility, and mitogenic activity. Some of the constituents of intermediate molecular weight are found to be highly active in microgram amounts in one assay while being virtually inactive in the other two, thus showing high specificity in their effects on cells. Current work suggests that the day may not be too far off when serum can be replaced by a small supplement of a few purified serum constituents. The advantages of a fully defined medium are obvious; in addition there is an important advantage to be gained from not having to use whole serum, and this derives from the fact that the elimination of microbial contaminants, especially mycoplasmal contaminants, has never been entirely satisfactory.

DENSITY DEPENDENT GROWTH

Growth of microorganisms in culture is regulated by the supply of nutrients; if this simple situation pertained in fibroblast cultures, there would be little more to be said. It is easy, however, to show that the situation is not so simple. Where fibroblast cultures are initiated at a low density and maintained by changing the medium at regular intervals, the cells initially proliferate rapidly, but this rate of growth is not maintained, it slows, and eventually the population plateaus. The plateau could represent a balance between cell loss and cell replenishment masking a significant rate of growth. This is not usually the case however, and the plateau indicates a quiescent culture within which proliferation has all but ceased. However, the medium taken off such quiescent cultures can be used to grow more fibroblasts, and in certain situations indeed, such as initiating colonies by clonal growth, it grows cells better than fresh medium. Thus not only is the medium from quiescent cultures rich enough in nutrients to support the growth of cells at lower densities, it is also enriched in products that protect and stimulate cells set out sparsely. These observations show that fibroblast growth is regulated in some way by the density of the cells. This is confirmed by the observation that a fixed aliquot of medium can yield a hundredfold or greater harvest of new cells from a thin culture than from a thick one.

One often reads in the literature statements to the effect that normal fibroblasts become density inhibited at confluence. Such statements are misleading and are to be deprecated because they hide an interesting aspect of the complex phenomenon of density inhibition. It is true that the much employed adult human skin fibroblast conforms to this generalization; so too does the popular murine 3T3 cell, but fibroblastic lines from some other human sources certainly do not. For example in our laboratory, using 10% serum, we expect to count about 2×10^6 cells in a 50 mm plate of adult human skin fibroblasts, and a just confluent culture yields around $1.5–2.0 \times 10^6$ cells. On the other hand, our stationary cultures of human fetal lung fibroblasts yield counts of $10–12 \times 10^6$ cells, and kidney and salivary gland fibroblasts intermediate counts between 4 and 6×10^6 cells. Density inhibition is not therefore always synonymous with confluence, and the existence of lines showing impeccable density inhibition, which under standard conditions will reliably discriminate in their behaviour between different degrees of over confluence, provide a challenging dimension to the problem which is almost completely ignored under the current preoccupation with the 3T3 cell.

The phenomenon we are dealing with has been variously called contact inhibition of growth (Levine *et al.*, 1965; Eagle and Levine, 1967), post-confluence inhibition (Martz and Steinberg, 1972), and density dependent inhibition of growth (Stoker and Rubin, 1967; Oren and Kohn, 1969).

I propose to discuss the phenomenon of density dependent inhibition of growth, without saying much about the molecular mechanisms involved. Not all cultured cells exhibit density dependent growth. Many virus transformed lines, and lines originating from malignancies grow independently of density, and those cells that can be grown in suspension culture approximate more to microorganisms in their behaviour. Nevertheless, all human diploid fibroblastic lines, to my knowledge, are characterized by density dependent growth and anyone growing normal fibroblasts will encounter the phenomenon.

MECHANISMS OF GROWTH REGULATION

Conditions under which density inhibition can be avoided or partially overcome

To explore the phenomenon further, let us note first one situation in which density dependent growth can be avoided, and second, the various ways in which growth inhibition can be partially overcome. Cultures continuously perfused with fresh medium achieve very high population densities (Kruse and Miedema, 1965; Kruse *et al.*, 1969). The plateau eventually established is really a pseudo-plateau resulting from the loss of poorly attached cells from the top of the sheet in the presence of continued growth. It appears therefore that the standard way of culturing cells under intermittent medium change favours density dependent growth and the results from perfusion experiments raise the question whether we can ignore partial medium depletion as a factor in growth regulation; I shall come back to this point later. It is interesting that where cultures are transferred to a regimen of continuous fresh medium perfusion, after the cells have ceased growing under intermittent medium change, renewed, unlimited growth is not reliably obtained and a lesser stimulation usually results.

A similar partial stimulation of quiescent cells can be obtained in many ways and the list of effective procedures grows every year. I will refer to wounding of the cell sheet later within the context of cellular interactions. There are reports in the literature of a greater or lesser stimulation of mitosis among quiescent cells by serum factors, proteases, hyaluronidases, RNAses, detergents, digitonin, insulin, colcemid and vinblastine (Vasiliev *et al.*, 1971). It is going too far, no doubt, to suggest that stimulation can be obtained by throwing anything onto quiescent cells, but given such a heterogeneous assortment of so-called stimulators one can be forgiven for wondering whether density dependent growth is sensitive to any change that creates a disturbance in the culture. The fact that density dependent growth can be partially and transiently upset by many factors should make us wary of accepting simplistic explanations for what is surely a complex phenomenon, the resultant of many influences acting on the cells.

Perhaps the paradigm example of a small, limited and transient stimulation of quiescent cells is provided by the routine, intermittent medium change (Bard and Elsdale, 1971). Two days following a medium change a stationary culture of fibroblasts observed by time-lapse microcinemaphotography reveals virtually no cell division and virtually immobile cells—quiescence indeed! Within half an hour of a medium change there is a striking and often drastic resurgence of cell movement, so violent may be the activity that temporary rents appear in the cell sheet. This activity dies down in the course of a day and is followed by division of 2–6% of the cell population.

Medium depletion

The last observation serves as a convenient reintroduction to the question of medium depletion in relation to density dependent growth. Although I have been at pains to point out that growth regulation in fibroblast cultures cannot be explained as a simple response to nutritional insufficiency, nevertheless there is ample evidence that medium depletion and the inability of cells to extract enough from the medium play a role. Indeed, the fashionable view at the moment, largely derived from work on 3T3 cells and other established lines, stresses the importance of serum factors for growth and sees density inhibition as a reflection of the fact that confluent cells prevent one another from obtaining enough serum mitogens. In effect, dense cells are starved of serum factors even though there may be plenty in the medium. A similar stimulation of a minority of human diploid fibroblasts in a quiescent culture to that obtained following a complete change of medium is also obtained after the addition of glucose and amino acids to the old medium; the addition of fresh serum alone to old medium has very little stimulatory effect (Smets, 1971). The former stimulation is preceded by a resurgence of cell movement, the latter non-stimulation is not, suggesting a correlation between movement resurgence and subsequent growth (Elsdale and Bard, 1972). Griffiths has correlated the fall in intracellular amino acids, and protein synthesis concon...tant with quiescence with depletion of essential amino acids in the medium (Griffiths, 1970). It is also observed that the response of quiescent cells to glucose and amino acid supplementation is itself density dependent, the higher the cell density the less the response (Ellem and Micronescu, 1974).

These observations have led people to conjecture that quiescent cells are less efficient users of medium nutrients than cells proliferating at lower densities. Notice that if this were so, it does not necessarily explain how cells become density inhibited. Furthermore, this suggested inefficiency could result from a number of causes. Dense cells may have less exchanging surface, their exchange rates may be decreased, again for a variety of reasons and dense cells may require to take on more nutrients than lower density cells to perform the same activities. An investigation by Michael Stoker is relevant here (Stoker, 1973). He surmised on physical chemical grounds that each cell in a culture is immediately surrounded by a stationary layer of fluid across which steep gradients in nutrient concentrations become established. As a result the cell effectively draws nutrients from a solution more dilute than the bulk medium. This layer is called a boundary layer. The denser the cells, the more pronounced the boundary layer effect and thus the lower the effective exchange rates. The depth of the boundary layer is dependent on a number of factors including the rate of flow in the ambient fluid. To test the hypothesis that the boundary layer was a growth limiting factor, Stoker arranged for a recycling of the medium over cultures, the medium being returned via a stationary capillary positioned above the cell sheet. Notice that this is different from continuous perfusion with fresh medium, since here a fixed aliquot of medium is being recycled. That portion of the cell sheet beneath the capillary was exposed to a continuous rapid flow of medium that fell off to the sides. In this way Stoker contrived to effectively blow away the boundary layer over a portion of the cell sheet and he observed a striking differential growth in response. These experiments provide good evidence for the existence of a boundary layer effect important for growth regulation. The only proviso with these experiments is that it appears that rather vigorous flow rates are required to cause density inhibited cells to resume growth. It is not excluded that the shear forces so generated may have direct effects, on the cell membrane perhaps, disturbing cell contacts.

Stoker interprets his results to support the conclusion that density dependent growth results from the inability of dense cells to extract from the medium enough of the things they need to grow. However, using an oscillating platform to diminish diffusion boundary effects in cultures of human diploid fibroblasts Froehlich and Anastasslades (1975) arrive at a different conclusion. Their results show that diffusion gradients may determine the rate at which cells grow but they do not affect the density at which growth ceases. They conclude that 'specific cell to cell interactions might still have to be invoked to explain density dependent inhibition'.

Numerous investigations show that quiescent growing cells incorporate labelled metabolites at different rates. For example, the rates of phosphate and uridine incorporation decrease markedly as cells become density inhibited (Webber, 1971). However, in the absence of controls on the boundary layer effect it is not always clear how these results should be interpreted.

Contact interactions and density dependent inhibition

Although nett growth does not cease until confluence is attained or thereafter, the rate of cell proliferation declines with increasing density in preconfluent cultures. The decline in growth rate culminating in quiescence is smooth and data from cell counts give no evidence for a stepped process. The decline is measurable at densities as low as 1/7th confluence and correlates with the establishment of significant numbers of contacts between cells and the progressive increase in contacts (Bard and Elsdale, unpublished). Similar observations have been made on 3T3 cells (Canagaratna and Riley, 1975). These observations suggest inhibition resulting from short range interactions. The seeming importance of contact is demonstrated by wounding experiments whereby patches of the quiescent cell sheet are removed (Todaro *et al.*, 1965; Dulbecco and Stoker, 1970; Raff and Houck, 1970). The cells at the edge of the wound migrate into the empty region, divide there and the hole is repaired, whereas neighbouring cells not at the wound edge but bathed in the same medium remain non-proliferating. It is clear from these experiments that the release from contact, or at least a reduction in the area of contact with neighbouring cells is a precedent for cell division, but serum factors are important too (Dulbecco, 1970). The current interpretation is that the cells at the wound edge divide because they obtain more serum.

Another experiment further suggests the importance of contact in density dependent inhibition and also reveals that an element of cellular specificity is involved (Bard and Elsdale, 1971). Use was made of adult human skin fibroblasts growing to low stationary densities of $1-1\frac{1}{2}$ monolayer equivalents, and human fetal lung fibroblasts growing to high stationary densities of 6–8 monolayer equivalents. Dishes were initiated with low density cells and maintained throughout on a complete medium change every second day. The low density cells were grown until stationary. A circle of 13 mm diameter was scraped clean of cells in each dish. A small inoculum of high density cells was added with the next medium change. Some of these cells fell onto the stationary lawns of low density cells and incorporated therein, whilst some fell onto the scraped area and initiated there a pure population of high density cells. The circumference of the 13 mm circle was scraped daily to prevent cell contacts across it. When all nett growth in the cultures had ceased, cell counts were made of the pure high density cell population and the surrounding mixed population, and compared with appropriate control cultures. The results showed that the high density cells in the mixed population could not have divided more than once (there is a probable connection here with the 'overshoot' phenomenon discussed by Martz and Steinberg (1972). The contact isolated, pure population of high density cells grew to the same high stationary density as the appropriate controls. The experiment shows that an excess of stationary low density cells can inhibit the growth of high density cells but, under the conditions of our experiment, only via contact. There was no evidence for a medium-borne inhibition. Inhibition of high density cells by low density cells has been observed in lines of non-human origin by Weiss and Njeuma (1971) and Todaro *et al.* (1965) among others.

There is a further aspect to contact inhibiting interactions, and it is one that has received little attention. In the ordinary way the cells in stationary cultures of fibroblasts are not evenly distributed. Local multilayering occurs, and in the case of high density cells such as fetal lung fibroblasts the contours are markedly uneven, and thick multilayered ridges are separated by valleys one cell thick. We know that the multilayers form as a result of cellular translocations and not as a result of differential growth *in situ* (Elsdale and Bard, 1972; Elsdale and Foley, 1969). Both in the case of normal growth towards quiescence, and in the case of reversals of density inhibition, it has been observed that mitoses are more frequent in the valleys than in the heaps (Ellem and Mironescu, 1972). In passing we note that this suggests that contact inhibition of growth, if it occurs, would be not an all or none response but a graded response. It also suggests that cells are responding solely to the density of cells in their immediate neighbourhood. If this is the case then the culture as a whole is to be equated with the summation of all local neighbourhoods. However, we have been able to contrive an experiment that casts doubt on whether the postulate of local density inhibition is the correct interpretation of the results just mentioned.

Under local density inhibition an uneven distribution should promote higher stationary densities than a uniform distribution. This follows because the sequestration of some of the cells, along with their inhibitory potential, into heaps, should leave the remainder freer to continue growing. This should be so whether the cells inhibit one another by contact, or whether they serve to reduce the availability of serum to their immediate neighbours. It is possible to culture fibroblasts uniformly distributed, and the trick is to herd them into a single parallel array. This is achieved by seeding the cells into dishes bearing a temporary carpet of aligned collagen bundles. The parallel array so established is perfectly stable, no ridges and valleys develop, and the distribution remains uniform. We have counted stationary cultures of the same line of cells varying only the distribution of the cells, uniform as against nonuniform, and the counts are not significantly different. If this result is taken at its face value it suggests that the phenomenon is not one of response to local density, but one of response to overall density, implying an averaging of cell states over the culture as a whole.

Thus, although the current fashion is to deny the importance of contact interactions in growth regulation, there are four observations pertaining to human diploid fibroblasts that are not easily subsumed under the view that regulation is to be accounted for entirely in terms of medium depletion and serum factors:

(1) Density inhibition is significant well prior to confluence,

(2) The inhibition of high density cells by low density cells,

(3) Stationary density is independent of the uniform or nonuniform distribution of the cells,

(4) The phenomenon of post-confluent growth: The possibility of an averaging of cell states is not an academic matter because it leads directly into the problem of post confluent growth and regulation.

Post-confluent growth

A post-confluent culture can be schematized as three layers, bottom, middle and top. Bottom cells have plastic beneath them and cells on top of them. Middle cells have cells all round them, top cells have cells beneath them only. The immediate circumstances of the cells in the three layers are quite different and it is difficult to envisage all cells in a dense culture receiving equal encouragement to multiply. In particular top cells have preferential service at the food counter, and other things being equal they should be the most favoured. The schematization used here would of course be meaningless if there were much tooing and froing between the layers. In the case of normal fibroblasts grown in the ordinary way there can be very little such movement because the cells become organized in leaf-like parallel arrays with an orthogonal alternation between vertically adjacent leaves, and this structure would be destroyed by movement between the leaves. Inhibition of top cells must in some way depend on the thickness of the middle layer if the growth regulating mechanism is to discriminate between degrees of overconfluence. Hence the attractiveness of an hypothesis of an averaging of cell states.

It seems therefore that current ideas on growth regulation may be insufficient to account for the growth behaviour of human diploid fibroblasts. Although serum availability is undoubtedly essential for the growth of cells, it is debatable whether this is the only control operating in density dependent growth regulation. Four phenomena observed in cultures of human diploid fibroblasts are itemized on page 39, and taken together they suggest the need for a more sophisticated hypothesis. It may be necessary to invoke some form of contact inhibition of growth not as an alternative, but as an adjunct to serum availability. There is no mechanistic difficulty about contact inhibition of growth given the evidence for cellular communication among cultured cells (Furshpan and Potter, 1968; Subak-Sharpe et al., 1969).

The following six postulates for a 'message hypothesis' take into account the data discussed above, and provide an example of how contact inhibition might work within the context of a serum requirement.

(1) Each cell is a source and receiver of inhibiting messages, with the likelihood that quiescent cells contribute stronger messages than growing cells (autocatalytic contact of Martz and Steinberg (6)). Hence the level of inhibitory messages will increase as more cells become quiescent through serum deprivation.

(2) A cell receiving a strong enough message stops growing regardless of other factors.

(3) Messages are propagated from cell to cell via contact.

(4) In the course of propagation through the network of cell contacts, messages tend to become normalized, so that all cells are in receipt of roughly the same message.

(5) In order to account for the inhibition of high density cells by low density

cells, the latter are assumed to be sources of stronger messages.

(6) Inhibition of cells otherwise favourably placed to grow (top cells, for example) requires the continuous receipt of inhibitory messages. Even a short interruption of message reception may allow a cell to escape into S phase (as might occur by the breaking of cell contacts due to a stimulation of motility).

References

Bard, J. and Elsdale, T. (1971). Specific growth regulation in early subcultures of human diploid fibroblasts. *In*: G. E. W. Wolstenholme and J. Knight (eds.). *Ciba Foundation Symposium on Growth in Cell Culture*, p. 187. (London and Edinburgh: Churchill Livingstone)

Canagaratna, M. C. P. and Riley, P. A. (1975). The pattern of density dependent growth inhibition in murine fibroblasts. *J. Cell. Physiol.*, **85**, 271

Dulbecco, R. (1970). Topoinhibition and serum requirement of transformed and untransformed cells. *Nature (London)*, **227**, 802

Dulbecco, R. and Stoker, M. G. P. (1970). Conditions determining the initiation of DNA synthesis in 3T3 cells. *Proc. Nat. Acad. Sci. U.S.A.*, **66**, 204

Eagle, H. and Levine, E. M. (1967). Growth regulatory effects of cellular interactions. *Nature (London)*, **213**, 1102

Ellem, K. A. O. and Mironescu, S. (1972). Mechanism of regulation of fibroblastic cell replication. I. Properties of the system. *J. Cell. Physiol.*, **79**, 389

Ellem, K. A. O. and Micronescu, S. (1974). The mechanism of regulation of fibroblastic cell replication. III. A central role for nutrient limitation: a conditioning effect due to serum imprinting of the cell response. *Exp. Cell Res.*, **88**, 175

Elsdale, T. and Bard, J. (1972). Cellular interactions in mass cultures of human diploid fibroblasts. *Nature (London)*, **236**, 152

Elsdale, T. and Foley, R. (1969). Morphogenetic aspects of multilayering in petri dish cultures of human fetal lung fibroblasts. *J. Cell Biol.*, **41**, 298

Froehlich, J. E. and Anastassiades, J. P. (1975). Possible limitation of growth in human fibroblast cultures by diffusion. *J. Cell. Physiol.*, **86**, 567

Furshpan, E. J. and Potter, D. D. (1968). Low-resistance junctions between cells in embryos and tissue culture. *In*: A. A. Moscona and A. Monray (eds.). *Current Topics in Developmental Biology*, **3**, p. 95. (New York: Academic Press)

Griffiths, J. B. (1970). The quantitative utilization of amino acids and glucose and contact inhibition of growth in cultures of the human diploid cell, W1-38. *J. Cell Sci.*, **6**, 739

Ham, R. G., McKeehan, W. L., McKeehan, R. A., Hammond, S. L., Lemmon, B. J. and Shipley, G. D. (1976). Normal cells do not need large amounts of serum protein. *In vitro*, **12**, 303

Kruse, P. F. and Miedema, E. (1965). Production and characterization of multi-layered populations of animal cells. *J. Cell Biol.*, **27**, 273

Kruse, P. F., Whittle, W. and Miedema, E. (1969). Mitotic and non-mitotic multiple-layered perfusion cultures. *J. Cell Biol.*, **42**, 113

Levine, E. M., Becker, Y., Boone, C. W. and Eagle, H. (1965). Contact inhibition, macromolecular synthesis and polyribosomes in cultured human diploid fibroblasts. *Proc. Nat. Acad. Sci. U.S.A.*, **53**, 350

Martz, E. and Steinberg, M. S. (1972). The role of cell-cell contact in 'contact' inhibition of cell division: a review and new evidence. *J. Cell. Physiol.*, **79**, 189

Mohamed, N. W., Mercer, G. E. and Holmes, R. (1976). Growth of human cells in chemically defined medium. *In vitro*, **12**, 304

Oren, R. and Kohn, A. (1969). Density-dependent inhibition of cell growth in cultures of primary and established lines of cells. *J. Cell. Physiol.*, **79**, 189

Raff, E. C. and Houck, J. C. (1970). Migration and proliferation of diploid human fibroblasts following 'wounding' of confluent monolayers. *J. Cell. Physiol.*, **74,** 235

Smets, L. A. (1971). Medium depletion and contact inhibition of replication: absence of a specific inhibitor. *Cell Tissue Kinet.*, **4,** 233

Stoker, M. G. P. (1973). Role of diffusion boundary layer in contact inhibition of growth. *Nature (London)*, **246,** 200

Stoker, M. G. P. and Rubin, H. (1967). Density dependent inhibition of cell growth in culture. *Nature (London)*, **215,** 171

Subak-Sharpe, H., Burk, R. R. and Pitts, J. (1969). Metabolic co-operation between biochemically marked mammalian cells in tissue culture. *J. Cell Sci.*, **4,** 353

Temin, H., Pierson, R. W., Jr. and Dulak, N. C. (1972). The role of serum in the control of multiplication of avian and mammalian cells in culture. *In*: G. H. Rothblat and V. J. Cristofalo (eds.). *Growth, Nutrition and Metabolism of Cells in Culture*, pp. 50–75. (New York and London: Academic Press)

Todaro, G. J., Lazar, G. and Green, H. (1965). The initiation of cell division in contact-inhibited mammalian cell line. *J. Cell. Comp. Physiol.*, **66,** 325

Vasiliev, J. M., Gelfand, I. M. and Guelstein V. I. (1971). Initiation of DNA synthesis in cell cultures by colcemid. *Proc. Nat. Acad. Sci. U.S.A.* **68,** 977

Webber, M. J. (1971). Phosphate transport, nucleotide pools, and ribonucleic acid synthesis in growing and in density-inhibited 3T3 cells. *J. Biol. Chem.*, **246,** 1828

Weiss, R. A. and Njeuma, D. (1971). Growth control between dissimilar cells in culture. *In*: G. E. W. Wolstenholme and J. Knight (eds.). *Ciba Foundation Symposium On Growth Control in Cell Cultures*, p. 169. (London and Edinburgh: Churchill Livingstone)

4

Cryopreservation of Tissue Culture Cells

D. E. PEGG

Division of Cryobiology, Clinical Research Centre, Northwick Park Hospital, Harrow, Middlesex

Other papers in this symposium have shown the value of tissue culture methods in the investigation of inherited metabolic disorders. For laboratory studies of this sort the ability to store cells carrying genetic defects is of obvious value. Defective cells would then be constantly available for reference purposes, for experimental use, and ultimately, one hopes, their normal counterparts for therapy. In this paper I propose to deal first with the basic principles of cryopreservation, then to consider specifically the storage of tissue culture cells, and finally to discuss the apparatus required for practical cell banking.

BASIC PRINCIPLES OF CRYOPRESERVATION

Cryopreservation may be defined as the storage of viable cells for long periods of time at temperatures significantly below the freezing point of physiological media. The vagueness of the definition is intentional for not all techniques involve freezing and in general the *cells* are not actually frozen. The storage temperature is not specified precisely but in practice is usually in the region of −196 °C although higher temperatures may, in some circumstances, be used. Cells may be effectively stored for many years, and in some circumstances virtually indefinitely. However, it is true that freezing usually occurs in cryopreservation techniques, and that freezing is normally lethal to cells unless some 'cryoprotective' agent is present. We will therefore consider in turn the occurrence of freezing in biological systems, the mode of action of cryoprotectants, and their application to cell preservation.

Freezing

Freezing is the separation of pure water as ice, which concentrates any solutes present in the remaining liquid phase. Cell damage from freezing may be caused by ice itself or by the altered liquid phase. Let us look first at the changes that occur in the liquid phase.

The principal solute in biological fluids is sodium chloride. When isotonic (0.15 M) sodium chloride solution is cooled it may supercool a few degrees, but if 'seeded' it freezes at $-0.56\,°C$. As cooling is continued so further ice separates, sufficient at each temperature to concentrate the salt in the remaining liquid and to produce a solution that has that freezing point. Thus, the remaining solution is progressively diminished in volume and increased in strength until at $-21.1\,°C$ the saline has reached a concentration of 5.2 M; at this temperature, the eutectic point, the remaining solution solidifies. Therefore, when cells are suspended in isotonic saline that is frozen, they are subjected to a 32-fold increase in sodium chloride concentration. In the mixed solute systems that occur in practice, similar changes in osmolality occur, but in addition there are changes in composition brought about by differing solubility characteristics of the various solutes. This may have important consequences. Van den Berg (1959) and van den Berg and Rose (1959) have shown for example that when a solution containing NaH_2PO_4 and Na_2HPO_4 is cooled, its pH falls if the molecular ratio of the two compounds is less than 57 but the actual pH is greatly influenced by the other solutes present. Mazur (Mazur *et al.*, 1969) has coined the term 'solution effects' to include all the changes that occur in the liquid phase as a result of freezing, together with the effects that they have on any cells present in the system.

It is important to realize that ice formation is normally entirely extracellular. There are several reasons for this. In the first place, when heat is removed by conduction from the external surface of the specimen, the coldest point will always be in the extracellular fluid. Secondly, the extracellular fluid forms one large compartment, whereas the intracellular space consists of very many small compartments. The probability of ice nucleation occurring in any given compartment is directly related to its size, and this makes it inevitable that nucleation will occur in the larger extracellular fluid compartment before a significant number of cells have begun to freeze internally. Once ice has started to form, it will propagate throughout that compartment until equilibrium is reached. Hence, even if a few cells should freeze internally before extracellular freezing starts, once extracellular ice has formed it will continue to grow in that space. Since cell membranes are impermeable to the main solutes present, water will be withdrawn from the cells by the increased external osmolality. Cells will shrink, and so long as cooling is slow enough to allow water to leave the cells and maintain equilibrium, no further intracellular freezing will occur. Let us look next therefore at the effects of raised external osmolality, and then return later to the possibility of intracellular freezing at rapid cooling rates.

Clearly, the most obvious effect of raised external osmolality is that cells shrink. Farrant and Woolgar (1972a; 1972b) measured the mass of water in erythrocytes exposed to solutions of sodium chloride and sucrose, and showed that the cells reached a minimum volume at about 1800–2000 mOsm/kg. Meryman (1968) suggested that when the osmolality is increased beyond this value, an actual hydrostatic pressure difference is produced across the cell membrane, and that this may then damage the cell. However, this seems distinctly unlikely, because intracellular proteins reach such a high concentration under these conditions, and their osmotic properties are so extremely non-ideal that very small water movements are sufficient to maintain equal water activity on either side of the cell membrane. Whatever the mechanism the membranes are damaged; they become leaky to cations and significant cell lysis is observed but this effect is not sufficient to explain the damage observed during freezing. Three observations made on human erythrocytes by Daw, Farrant and Morris (1973) illustrate that additional factors are at work during freezing. First, the amount of haemolysis produced at a given osmolality without freezing was less than that observed when the same osmolality was produced by freezing. Moreover there was less damage when cells were frozen in sucrose instead of sodium chloride, whereas the same shrinkage and haemolysis was produced by exposure to sucrose without freezing. Finally, freezing was found to render red cell membranes permeable to sucrose, whereas exposure to equivalent osmolalities without freezing did not.

The obvious differences between the simple hypertonic model system and actual freezing are that freezing and thawing involve changes in temperature as well as concentration, and that during thawing the cells are resuspended in the original osmolality. Lovelock in 1955 showed that erythrocytes suspended in hypertonic salt solutions sufficient to produce only minimal haemolysis if the temperature was held constant at 37 °C, were lysed when cooled to 0 °C. This is a 'thermal shock' phenomenon, possibly analogous to that which occurs naturally with some species of bacteria (Meynell, 1958) and spermatozoa (Smith, 1961) but which occurs with erythrocytes only when they have been appropriately sensitized. Lovelock (1955) showed that hypertonic conditions elute lecithin from erythrocyte membranes. He suggested that the low melting point of this lipid normally makes lecithin-rich membranes more pliable at low temperatures and therefore less liable to fail under the stresses produced by differential thermal contraction. During freezing red cells are exposed both to a falling temperature and to rising salt concentrations which may elute membrane lecithin.

Lovelock (1953a) also carried out a very elegant experiment showing that red cells exposed to hypertonic saline were damaged further when they were returned to isotonic conditions. He showed that when erythrocytes were frozen in 0.15 M sodium chloride, they could not be recovered without lysis if they spent more than 30 s in the temperature zone between −4 and −40 °C. He pointed out that −4 °C is the freezing point of 0.8 M NaCl, and that therefore as the cells are cooled through this zone they are exposed to external NaCl concentrations increasing from 0.8 M at −4 °C to 5.2 M at −21.1 °C. When erythrocytes were suspended in NaCl solutions varying from 0.6 to 3.6 M, the degree of haemolysis produced when they were resuspended in 0.15 M sodium chloride was the same as that which occurred when suspensions in isotonic saline were frozen to the temperature which results in the same concentration and then thawed again (see Figure 4.1). Lovelock (1953a) argued that during

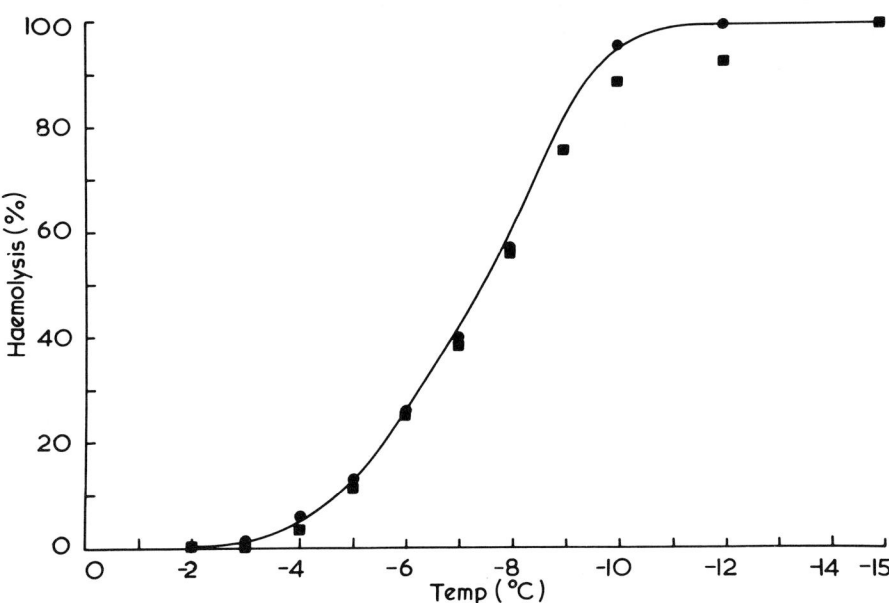

Figure 4.1 Haemolysis produced by exposing red cells suspended in 0.15 M NaCl solution to the temperatures indicated and rewarming them (●). The haemolysis produced by exposing red cells to solutions having the indicated freezing points, and then returning them to isotonic saline are also shown (■). The agreement between the data is apparent. (Data from Lovelock, 1953a) (Reproduced by permission of Blackwell Scientific Publications)

exposure to hypertonic saline the cells total cation content was increased above normal, so that on restoration to isotonic conditions the net cation loading led to osmotic lysis. Moreover, if the cells remained cation-leaky on thawing the internal impermeant solutes would lead to colloid osmotic haemolysis (Woolgar, 1972), thus providing a further mechanism of osmotic lysis.

Thus, we now look upon the damage occurring during slow cooling, when ice is exclusively extracellular, as being due to solution effects alone, and that this damage occurs in two stages. In the first stage cell membranes are damaged and the cells become loaded with additional solute but remain apparently intact. In the second stage the damage becomes manifest through the additional stresses of cooling (thermal shock) and resuspension in isotonic saline as a result of thawing (osmotic shock).

The discussion so far has been limited to slow cooling where ice is exclusively extracellular. Since the processes that damage the cells during cooling are physical in nature, and since the rate of such processes is temperature-dependent, it would be anticipated that accelerating the cooling rate would increase cell survival. Direct experiment has confirmed this expectation but has also shown the survival eventually falls off if the cooling rate is increased still further. Mazur (1963) has produced a very convincing explanation for this phenomenon. He pointed out that red cells, which have a high permeability for water, have a high optimum cooling rate, whereas yeast, which is relatively impermeable to water, survives better at lower cooling rates. From the permeability data it follows that yeast cells will tend to freeze internally at lower cooling rates than erythrocytes, and he was able to show that significant amounts of supercooled intracellular water start to occur in yeast at 10 °C/min, and in erythrocytes at 3000 °C/min. These happen to be the cooling rates giving maximum survival with each of these cells, and if one postulates that intracellular freezing is damaging, by some mechanism as yet unknown, then the actual effect of cooling rate on cell survival can be explained. There is now direct experimental confirmation, both from light microscopy (Diller *et al.*, 1972) and from electron microscopic examination of freeze-substituted material (Walter *et al.*, 1975) that intracellular freezing does occur during rapid cooling and is associated with cell destruction. Figure 4.2 shows the effect of cooling

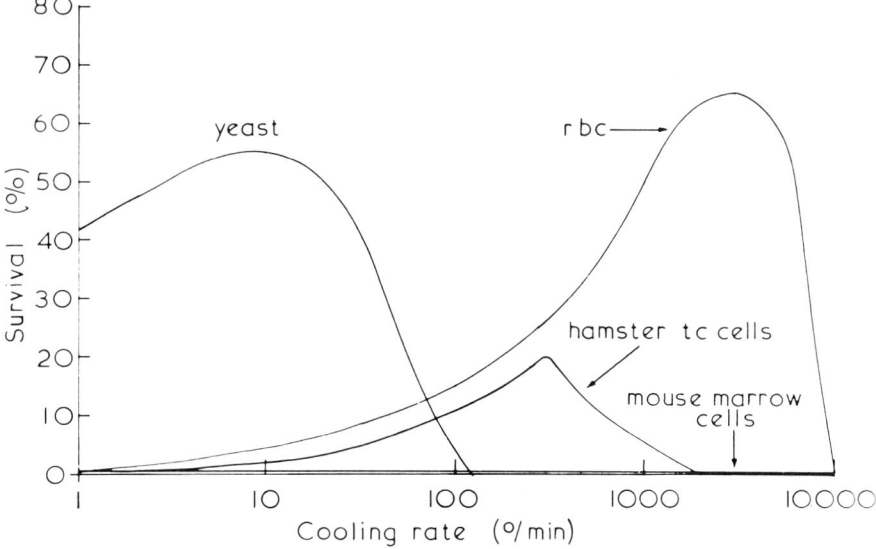

Figure 4.2 Graph showing the effect of cooling rate on the survival of yeast cells, tissue culture cells (hamster TC cells), human red blood cells, and mouse bone marrow cells. Note the differing optimum cooling rates, and the fact that less than 2% of the bone marrow cells survive at any cooling rate. (Reproduced by permission of IPC Science and Technology Press Ltd).

rate on the survival of four cell types; note particularly that mouse bone marrow cells, which are typical nucleated mammalian cells, show less than 2% survival at any cooling rate. A possible explanation is that accelerating the cooling rate causes internal freezing to occur before the rate is fast enough to reduce significantly the damage from solution effects. If this is correct, a useful survival rate would be obtained if some method could be found to reduce solution effects at cooling rates slow enough to avoid intracellular freezing. It is fortunate that the accidental discovery of the cryoprotective action of glycerol did exactly this.

Cryoprotection

The historic discovery of the cryoprotective effect of glycerol was made in 1948 when Polge, Smith and Parkes (1949) found that fowl spermatozoa that had been cooled to −79 °C in 1.1 M glycerol recovered with little damage after thawing. Mammalian erythrocytes behaved similarly. Lovelock (1953b) explained the cryoprotective action of glycerol on the basis of colligative action; he demonstrated that the rise in salt concentration when isotonic NaCl solution was frozen was much greater than the rise in glycerol concentration when biologically acceptable concentrations of glycerol were frozen (see Figure 4.3). When NaCl and glycerol are present, they will both be concentrated to the same proportional extent, since it is the removal of solvent water to form ice that is concentrating both solutes. The reduction in salt concentration achieved by the addition of glycerol to a suspension of erythrocytes in physiological saline, exactly paralleled the reduction in haemolysis when the suspension was frozen (Lovelock, 1953b). Whatever the glycerol concentration, haemolysis started when the NaCl concentration reached a mole fraction of 0.014 and reached 5% whenever the mole fraction of NaCl increased to 0.02, irrespective of temperature or glycerol concentration.

Nash (1962) studied many other neutral solutes that are cryoprotective for erythrocytes, and found that they had the common properties of ability to penetrate cells, lack of toxicity, and a high affinity for water. Clearly the last two characteristics are vital. Absence of intrinsic toxicity is a self-evident requirement, and the ability to produce concentrated solutions with a low freezing point is important because such compounds will be the most effective 'salt buffers'. However, the fact that many commonly used cryoprotective agents penetrate cell membranes actually *produces* some problems, and is certainly not a necessary property for cryoprotection.

Penetrating cryoprotective agents like glycerol and dimethylsulphoxide (DMSO) permeate a good deal more slowly than water and consequently they all produce osmotic transients, the severity and duration of which vary with the compound and the cell in question. In general, osmotic disturbances have more severe effects during thawing and resuspension of the cells in cryoprotectant-free medium (when the solute is leaving the cells) than during initial equilibration and cooling (when it is entering the cells). The important point is that if time is not allowed for penetrating cryoprotectants to leave the cells gradually during thawing and subsequent manipulations, then their presence may actually add to the osmotic damage suffered by stored cells. Recently, much more attention has been paid to the determination of optimum thawing procedures,

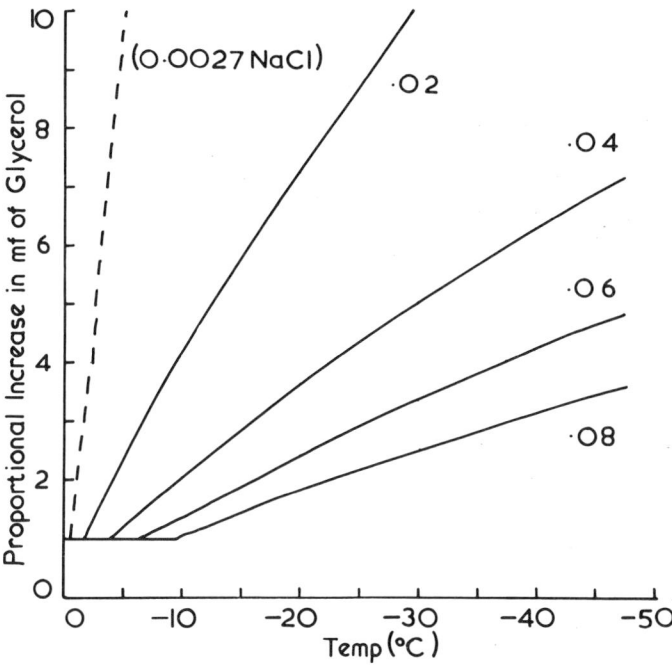

Figure 4.3 Graph showing the rise in concentration of the indicated initial mole fractions of glycerol (−) compared with isotonic sodium chloride (0.0027 mf; − − −). It is apparent that the increase in mole fraction of NaCl in a solution that also contains significant amounts of glycerol will be greatly reduced. (Reproduced by permission of Blackwell Scientific Publications)

and techniques for removing cryoprotectants (Rapatz *et al.*, 1975; Thorpe *et al.*, 1975; Mazur and Miller, personal communication).

For many years it was widely believed that thawing should be as rapid as possible in order to obtain maximum survival. Certainly there was considerable experimental support for this belief, and the explanation generally offered was that rapid thawing minimized the duration of exposure of cells to high concentrations of solutes at high temperatures. It is now clear however that there is an important interdependence of cooling and thawing rates, and that rapid thawing is advantageous only under certain conditions. It seems that in many of the empirically-derived successful preservation procedures, the cells do in fact contain very small ice nuclei which in themselves are harmless, but which grow during slow thawing, with lethal results. Rapid thawing is therefore beneficial. When other cooling rates are studied at a constant thawing rate, the familiar peak survival at one cooling rate is observed. At faster cooling rates gross intracellular freezing occurs and destroys the cells during cooling. At slow rates, the cells become more heavily loaded with solute, and therefore suffer osmotic lysis during thawing. However, when slower thawing rates are studied, it is found that high survival rates may be obtained at slow cooling rates (Whittingham *et al.*, 1972; Mazur and Miller, 1976). Such techniques do, of course, produce exposure to high solute concentrations at high temperatures, but it is precisely because they do this that they are able to reduce osmotic stresses. This interaction is now being energetically studied and will undoubtedly affect practical preservation techniques in the future.

A complementary approach to the general problem of solute loading and consequent osmotic damage, is the use of cryoprotectants that do not penetrate cell membranes, and therefore do not add to the difficulties produced by salt loading. In 1955 Bricka and Bessis showed that erythrocytes suspended in high concentrations of dextran or polyvinylpyrrolidone (PVP) would survive freezing at very lower temperatures, and it was subsequently shown that PVP would protect bone marrow and tissue culture cells (Mazur *et al.*, 1969). For many years it was supposed that such macromolecular compounds must have a different mechanism of action from glycerol, because the mole fraction of the protective compound was very low. However, the solution properties of macromolecules especially in high concentration are known to be non-ideal (Jellinek and Fok, 1967). Farrant and Woolgar (1969) have shown that the presence of 15–30% of PVP depresses the rise in salt concentration of the system $PVP–NaCl–H_2O$ sufficiently to enable red cells to survive exposure to $-10\,°C$ for 20 min. Even though these compounds do not penetrate into the cells they will reduce the build-up of salt concentration inside the cells, since intracellular solute concentration will be in equilibrium with the lower external salt concentration. An additional factor has been pointed out by Woolgar (1972) who showed that haemolysis was reduced by the presence of external colloid (PVP) when frozen red cells suspensions were thawed; this was probably due to the balance that the PVP provided for the intracellular colloid once the cell membranes had become leaky to cations. Other, more specific actions of certain high molecular weight materials may also play a part in cryoprotection. However, it seems likely that the most important action of high molecular weight, non-penetrating cryoprotective agents is the same as that proposed by Lovelock for glycerol.

It will be appreciated that all these cryoprotective agents affect only the part of the graph relating damage to cooling rate which has been ascribed to solution effects. This was elegantly demonstrated by Leibo and colleagues (1969) who showed for mouse bone marrow cells, that increasing glycerol concentration produced progressively greater surviving fractions at progressively lower cooling rates, but left the 'intracellular freezing' part of the curve completely unaffected, the survival being the same for all glycerol concentrations at high cooling rates (see Figure 4.4).

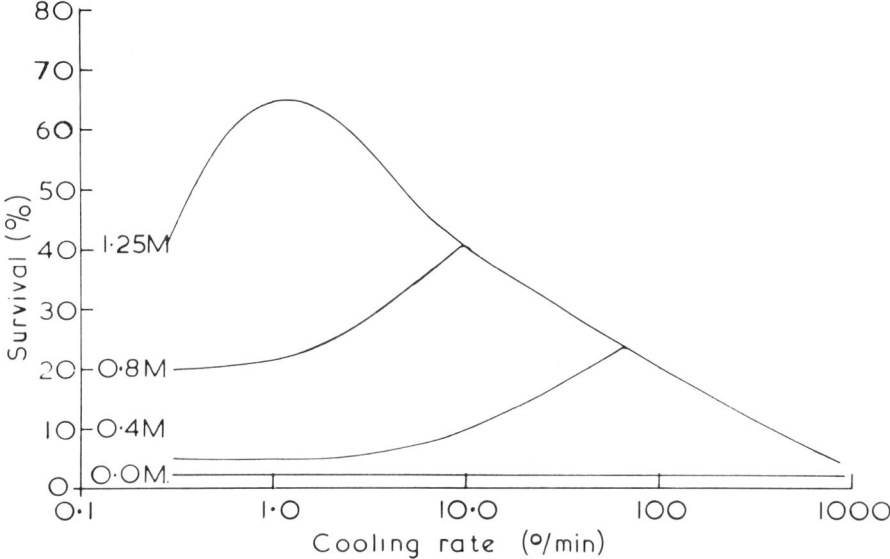

Figure 4.4 Graph showing the effect of cooling rate on the survival of mouse haemopoietic cells cooled in the presence of various concentrations of glycerol. Note that increasing the glycerol concentration moved the optimal cooling rate downwards and increased maximum survival. (Data from Leibo *et al.*, 1969). (Reproduced by permission of IPC Science and Technology Press Ltd)

Cell preservation

Long before much of the fundamental information reviewed above was available, experimental investigation had shown that many cells could be preserved by fairly slow ($\sim 1\,°C\ min^{-1}$) cooling in the presence of moderate (1–2 molar) concentrations of glycerol or DMSO, followed by rapid ($\sim 200\,°C\ min^{-1}$) thawing, and this came to be regarded as the 'standard technique'. The accumulation of information on the mode of action of cryoprotectants and the effects of cooling and thawing rates has shown that this is far too simple a view. Each cell type should be considered separately since variation of the cooling rate, the cryoprotective additive, and the technique of thawing and post-thaw handling will produce very different optimum conditions for the storage of each cell type. In the present state of knowledge there is no certain way of predicting the optimum method for an as yet untested cell type, but it is hoped that this discussion of the principles involved in cryopreservation will enable the reader to approach cell storage problems less empirically.

PRESERVATION OF TISSUE CULTURE CELLS

This 'standard technique' has been found to be effective for many tissue culture cell lines, for instance HeLa and L cells have been stored for five years by cooling at approximately 1 °C min^{-1} in serum with 0.65 M glycerol (for L cells) or 2.75 M glycerol (for HeLa cells) and holding at −79 °C (Scherer, 1960). Porterfield and Ashwood-Smith (1962) found 1.4 M DMSO to be superior to 1.4 M glycerol for chick fibroblasts and human embryo lung cells. These findings were confirmed by Dougherty (1962) who found 1 °C min^{-1} to be the optimum cooling rate and also demonstrated the value of including serum in the medium. Thawing was carried out rapidly by immersion in a 37 °C water bath and the cryoprotectant was removed by dilution with growth medium over a period of about 5 min. In our Institute fibroblasts from patients with metachromatic leukodystrophy and fibroblasts and amniotic cells from Hurler's syndrome, Hunter's syndrome and Lesch–Nyhan syndrome are all preserved by cooling at 1 °C min^{-1} in the presence of 0.6 M DMSO and 15–30% serum. The cells are stored at −196 °C and thawing is carried out rapidly in a 37 °C water bath. Dilution is then carried out with approximately three volumes of growth medium and the cells are plated out. One per cent DMSO does not inhibit cell attachment, and when the medium is changed on the following day, the cultures grow rapidly (Gibbs, personal communication). Such methods are of wide, but not universal applicability. For example, Chinese hamster lung fibroblasts gave very low survival rates when cooled at 1 °C min^{-1}. Survival rose to 60% when they were cooled in 1.25 M glycerol at 100 °C min^{-1} (Mazur et al., 1969) and rapid thawing was vital. Good survival at 10 °C min^{-1} was obtained when 15% w/v PVP (MW 40 000) was used as the cryoprotectant. This study underlines the importance of determining the optimum cooling rate and studying more than one cryoprotectant when faced with a new cell line.

Recently there has been a resurgence of interest in a somewhat different technique of cryopreservation known as two-step cooling. This method was originally developed by Luyet and Keane (1955) who found that spermatozoa suspended in a medium containing a suitable cryoprotective agent, could be effectively preserved if they were cooled rapidly to a critical subzero temperature, held at that temperature for a sufficient length of time, and then cooled rapidly to the storage temperature, usually −196 °C. Studies recently carried out by Walter, Knight and Farrant (1975) have shown that during the holding time, when the cells are exposed at a low temperature to a high external osmolality, they shrink, and that when they are then cooled rapidly, they do not freeze internally. Thus, the mechanism of cryoprotection is precisely the same as that in controlled-rate cooling. So far, the two-step method has been shown to be applicable to bull spermatozoa (Luyet and Keane, 1955), human lymphocytes and hamster fibroblastic tissue culture cells (Farrant *et al.*, 1974). A holding temperature of −25 °C for 10 min seems suitable and after rapid thawing, 90% survival of the fibroblasts was obtained under optimum conditions. There is an important interaction between holding temperature and holding time (see Figure 4.5). Two-step cooling will probably be applicable to

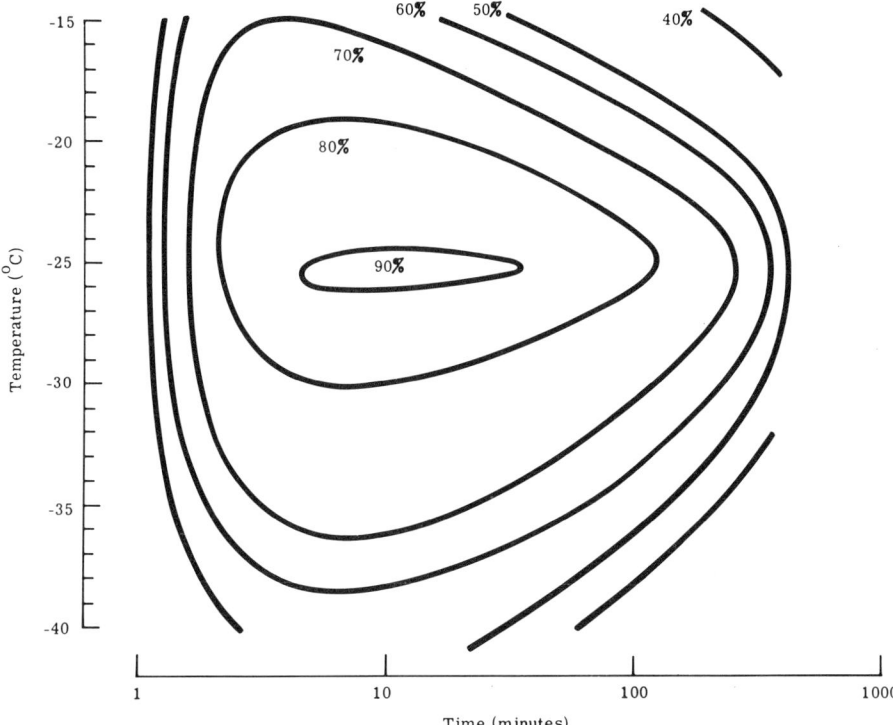

Figure 4.5 Contours of percentage survival of Chinese hamster lung fibroblasts held at different subzero temperatures for various times before rapid cooling to −196 °C and thawing. Cells were suspended in growth medium containing 10% fetal calf serum with 0.7 M DMSO. (Data and figure provided by L. E. McGann and J. Farrant)

other cell lines, although survival rate tends to fall off when the sample volume is increased above 1 ml due to the importance of obtaining rapid cooling and rewarming rates in this technique.

EQUIPMENT FOR CRYOPRESERVATION

Throughout this chapter emphasis has frequently been laid on the importance of cooling rate, of storage temperature, and during two-step techniques, of the intermediate holding temperature. Several sophisticated controlled-rate cooling machines have been described (Pegg *et al.*, 1973; Hayes *et al.*, 1974) and some are commercially available. They are convenient in use and provide a record of the cooling curve actually obtained, which is a valuable safeguard. All the machines currently on the market in the United Kingdom use liquid nitrogen as the refrigerant. One point of particular importance in practice is the control of supercooling. If unfrozen samples are placed in a machine and allowed to freeze spontaneously during cooling some freeze at their freezing point, while others supercool to a variable extent. When freezing occurs the sample temperature rises to its freezing point and when the latent heat has been dissipated, subsequent cooling is much faster because the environment has continued to cool. Up to 14 °C of supercooling has been observed. Cells that are very sensitive to more rapid cooling may be destroyed in this post-freezing phase. With many cells, this is not a significant problem, but with some, for example mouse embryos, it is *very* important (Leibo, 1976). The simplest way to avoid this difficulty is to seed the samples before controlled cooling is started. This is most easily done by placing the tubes in a bath about 2 °C colder than the sample freezing point (e.g. −7 °C for 1.4 M DMSO) and then tapping them vigorously after 2 minutes' incubation.

When selecting a controlled-rate cooler particular attention should be given to the programming system employed. Differential thermocouple controllers have serious drawbacks which are discussed elsewhere (Pegg *et al.*, 1973) and are not recommended, but cam-followers and other forms of direct programming are satisfactory. However, it is possible to build a simple cooling apparatus in the laboratory, and devices that rely on the immersion of an insulated cooling vessel in a bath of liquid nitrogen or methylated spirit cooled with solid carbon dioxide are perfectly practicable. Rather less satisfactory are the insulated hollow plugs designed to produce a slower cooling when inserted into the neck of a liquid nitrogen tank; the cooling rates obtained are quite variable and very dependent on the load of samples being cooled.

The constant-temperature bath, typically at −25 °C, required for two-step cooling is best provided by placing a bath of silicone oil or glycerol solution in an electrical refrigerator set to that temperature.

Storage of cells is most conveniently and reliably achieved by the use of liquid nitrogen refrigerators which provide temperatures of −160 °C to −196 °C. There is a wide choice of refrigerators on the market offering liquid or gas-phase storage or both, in sizes varying from some 7 litres to 640 litres capacity, and with or without elaborate inventory control systems. In making a choice between liquid and gas phase systems, the danger of samples exploding during thawing as a result of leakage of liquid nitrogen during storage should not be overlooked. With modern vacuum insulation techniques the temperature provided by gas-phase systems is quite low enough (<−160 °C) for all practical purposes. Very small and unspillable refrigerators are also available for the transport of frozen samples.

ACKNOWLEDGMENT

This chapter is based upon a paper previously published in the *Journal of Clinical Pathology,* volume 29, pages 271–285, 1976. The permission of the Editor and the British Medical Association to make extensive quotations from that paper is gratefully acknowledged.

References

Berg, van den L. (1959). The effect of addition of sodium and potassium chloride to the reciprocal system: KH_2PO_4–Na_2HPO_4–H_2O on pH and composition during freezing. *Arch. Biochem.,* **84,** 305

Berg, van den L. and Rose, D. (1959). The effect of freezing on the pH and composition of sodium and potassium phosphate solutions: the reciprocal system KH_2PO_4–Na_2HPO_4–H_2O. *Arch. Biochem.,* **81,** 319

Bricka, M. and Bessis, M. (1955). Sur la conversation des érythrocytes par congélation à basse température en présence de polyvinylpyrrolidone et de dextron. *C.R. Soc. Biol. Paris,* **149,** 875

Daw, A., Farrant, J. and Morris, G. J. (1973). Membrane leakage of solutes after thermal shock or freezing. *Cryobiology,* **10,** 126

Diller, K. R., Cravalho, E. G. and Huggins, C. E. (1972). Intracellular freezing in biomaterials. *Cryobiology,* **9,** 429

Dougherty, R. M. (1962). Use of dimethyl sulphoxide for preservation of tissue culture cells by freezing. *Nature (London),* **193,** 550

Farrant, J., Knight, S. C., McGann, L. E. and O'Brien, J. (1974). Optimal recovery of lymphocytes and tissue culture cells following rapid cooling. *Nature (London),* **249,** 452

Farrant, J. and Woolgar, A. E. (1970). Possible relationships between the physical properties of solutions and cell damage during freezing. *In*: G. E. W. Wolstenholme and M. O'Connor (eds.) *The Frozen Cell* pp. 97–114, Ciba Symposium (London: J. & A. Churchill)

Farrant, J. and Woolgar, A. E. (1972a). Human red cells under hypertonic condition a model system for investigating freezing damage.1. Sodium chloride. *Cryobiology,* **9,** 9

Farrant, J. and Woolgar, A. E. (1972b). Human red cells under hypertonic conditions: a model system for investigating freezing damage. 2. Sucrose. *Cryobiology,* **9,** 16

Hayes, A. R., Pegg, D. E. and Kingston, R. E. (1974). A multirate small-volume cooling machine. *Cryobiology,* **11,** 371

Jellinek, H. H. G. and Fok, S. Y. (1967). Freezing of aqueous polyvinylpyrrolidone solutions. *Kolloidzeitschrift,* **220,** 122

Leibo, S. (1976). Personal communication

Leibo, S. P., Farrant, J., Mazur, P., Hanna, M. G. and Smith, L. H. (1969). Effects of freezing on marrow stem cell suspensions: interactions of cooling and warming rates in the presence of PVP, sucrose or glycerol. *Cryobiology,* **6,** 315

Lovelock, J. E. (1953a). The haemolysis of human red blood cells by freezing and thawing. *Biochim. Biophys. Acta,* **10,** 414

Lovelock, J. E. (1953b). The mechanism of the protective action of glycerol against haemolysis by freezing and thawing. *Biochim. Biophys. Acta,* **11,** 28

Lovelock, J. E. (1955). Haemolysis by thermal shock. *Br. J. Haematol.,* **1,** 117

Luyet, B. J. and Keane, J. (1955). A critical temperature range apparently characterised by sensitivity of bull semen to high freezing velocity. *Biodynamica,* **7,** 281

Mazur, P. (1963). Kinetics of water loss from cells at subzero temperatures and the likelihood of intracellular freezing. *J. Gen. Physiol.,* **47,** 347

Mazur, P., Farrant, J., Leibo, S. P. and Chu, E. H. Y. (1969). Survival of hamster tissue culture cells after freezing and thawing: interactions between protective solutes and cooling and warming rates. *Cryobiology*, **6,** 1

Mazur, P., Leibo, S. P., Farrant, J., Chu, E. H. Y., Hanna, M. G. and Smith, L. H. (1969). Interactions of cooling rate, warming rate, and protective additive on the survival of frozen mammalian cells. *In*: G. E. W. Wolstenholme and M. O'Connor (eds.) *The Frozen Cell* pp. 69–85, Ciba Symposium (London: J. & A. Churchill)

Meryman, H. T. (1968). Modified model for the mechanism of freezing injury in erythrocytes. *Nature (London)*, **218,** 333

Maynell, G. C. (1958). The effect of sudden chilling on *Escherichia coli. J. Gen. Microbiol.*, **19,** 380

Nash, T. (1962). The chemical constitution of compounds which protect erythrocytes against freezing damage. *J. Gen. Physiol.*, **46,** 167

Pegg, D. E., Hayes, A. R. and Kingston, R. E. (1973). Cooling equipment for use in cryopreservation. *Cryobiology*, **10,** 271

Polge, C., Smith, A. U. and Parkes, A. S. (1949). Revival of spermatozoa after vitrification and dehydration at low temperatures. *Nature (London)*, **164,** 666

Porterfield, J. S. and Ashwood-Smith, M. J. (1962). Preservation of cells in tissue culture by glycerol and dimethyl sulphoxide. *Nature (London)*, **193,** 548

Rapatz, G., Luyet, B. and MacKenzie, A. (1975). Effect of cooling and rewarming rates on glycerolated human erythrocytes. *Cryobiology*, **12,** 293

Scherer, W. F. (1960). Effects of freezing speed and glycerol diluent on 4–5 year survival of HeLa and L cells. *Exp. Cell Res.*, **19,** 175

Smith, A. U. (1961). *Biological Effects of Freezing and Supercooling.* p. 440. (London: Edward Arnold)

Thorpe, P. E., Knight, S. C. and Farrant, J. (1976). Optimal conditions for the preservation of mouse lymph node cells in liquid nitrogen using cooling rate techniques. *Cryobiology*, **13,** 126

Walter, C. A., Knight, S. C. and Farrant, J. (1975). Ultrastructural appearances of freeze-substituted lymphocytes frozen by interrupting rapid cooling with a period at −26 °C. *Cryobiology*, **12,** 103

Whittingham, D. G., Leibo, S. P. and Mazur, P. (1972). Survival of mouse embryos frozen to −196 °C and −269 °C. *Science*, **178,** 411

Woolgar, A. E. (1972). A study of the effects of freezing on human red blood cells. (Ph.D. Thesis, Council for National Academic Awards)

5
Tissue Culture in the Study of Neuromuscular Diseases

V. DUBOWITZ

Department of Paediatrics and Neonatal Medicine (Institute of Child Health) and Jerry Lewis Muscle Research Centre, Hammersmith Hospital, London

The cultured fibroblast is now well established as a useful source for the identification of inborn errors of metabolism. The myoblast, on the other hand, has perhaps not yet quite arrived. This is probably not the fault of the myoblast, but merely reflects our current ignorance in relation to the fundamental biochemical lesion in most muscle diseases. The picture in relation to some of the metabolic myopathies is, however, becoming clearer.

ANIMAL MUSCLE IN CULTURE

Muscle explants can readily be grown in culture and will achieve a remarkable degree of differentiation. After a period of quiescence (lag phase), there is a proliferation of mononucleate myoblasts which undergo mitotic division. This is followed by fusion of cells to form syncytial multinucleate myoblasts, which elongate into myotubes with chains of central nuclei. These myotubes develop striated myofibrils and may show spontaneous contractions in isolated culture. Much of the fundamental work on myogenesis in culture has been done on embryonic vertebrate musculature, particularly the chick (Holtzer *et al.*, 1973), rat (Yaffe, 1971) and quail (Lipton and Konigsberg, 1972). There are, however, still large gaps in our understanding of some fundamental questions such as the cellular events which lead to fusion of cells.

HUMAN MUSCLE IN CULTURE

In contrast to the culture of animal muscle, which goes back more than half a century, it is only in recent years that attention has been focused on human muscle (Pogogeff and Murray, 1946). Even in the mid-1960s there was still considerable pessimism about the prospects of growing mature human muscle in culture, although a few workers had already had some degree of success, both with normal and dystrophic muscle. A number of these early authors reported abnormalities in dystrophic muscle in culture.

Geiger and Garvin (1957) found that although normal and dystrophic muscle initially appeared to grow in a similar way, the cells of the dystrophic muscle were more heavily granulated and failed to develop the cross-striations seen in normal muscle. After four weeks, cells from the dystrophic muscle hypertrophied, reaching lengths up to 3000 μm, and then degenerated. When subcultured the dystrophic muscle only survived one passage and then degenerated, whereas normal muscle could be maintained for as long as a year through as many as 12 passages.

O'Steen (1962) grew human muscle in diffusion chambers inserted into the peritoneal cavity of mice. He observed that in cultures from dystrophic muscle the size of mononucleate myoblasts was similar to that of normal cultures, whereas the multinucleate myoblasts were larger and the myotubes shorter. In a study of cultures from four normal and three dystrophic muscles, Goyle et al. (1967) found that myoblasts from dystrophic muscle were smaller and more variable in size and shape than the normal, and that there were fewer multinucleate cells. Kakulas and colleagues (1968) studied the primary explants of normal and dystrophic human muscle by time-lapse cinemicrophotography and observed differences both in the morphology and in the pattern of growth of dystrophic cells.

There was also some divergence in observations on the lag period between the initiation of the culture and the first cell growth phase. Thus, Geiger and Garvin (1957) observed a lag of two to six days in dystrophic cultures, compared to seven to 20 days in the normal, while Goyle and co-workers (1967) found a lag phase of three to five days in both normal and dystrophic cultures and O'Steen (1962) a lag phase of two to four days in both. Herrmann (1958) also found a similar lag phase in both normal and dystrophic muscle, and Herrmann and co-workers (1960) observed no striking difference between normal and dystrophic muscle in either the number of cultures showing outgrowth after one month or the maximal duration of in vitro maintenance of the culture.

A number of workers observed cross-striations in normal human muscle grown in vitro (Pogogeff and Murray, 1946; Geiger and Garvin, 1957; Herrmann et al., 1960; O'Steen, 1962) but not in dystrophic muscle (Geiger and Garvin, 1957; Herrmann et al., 1960; O'Steen, 1962). Goyle and co-workers (1968) claimed to have seen cross-striations in stained preparations from both normal and dystrophic muscle under polarized light, but the appearances illustrated were irregular and somewhat bizarre. Kakulas et al. (1968) noted the presence of multinucleated myotubes with cross-striations in the primary explants of dystrophic muscle after several days' growth.

Personal experience

In our initial study we found that diseased human muscle was capable of differentiation in culture in the same way as normal muscle and we were unable to detect any apparent differences in their morphological appearances (Skeate et al., 1969). Cross-striations were observed in a number of cultures from both normal and diseased muscle. There were no spontaneous contractions.

In a subsequent study of 81 biopsies from 54 patients with various neuromuscular disorders and 27 controls (Bishop et al., 1971), we obtained successful growth in 52 (64%). Once again no differences could be found between normal and diseased muscle in respect of various morphological parameters which we were able to quantitate, such as the length of the lag phase, the size of myotubes at specific times in culture, the number of nuclei, the incidence of myotubes and presence of cross-striations. During this study we were also able to freeze down trypsinized cultures for storage in liquid nitrogen and were subsequently able to successfully thaw out and regrow these cultures.

In an extension of this investigation, we subsequently demonstrated that there was also no distinction between normal and diseased muscle in culture with respect to histochemical enzyme reactions (Gallup et al., 1972a) or RNA and DNA synthesis studied by autoradiography (Gallup et al., 1972b).

In a recent further series of 63 biopsies, successful outgrowth of cells in culture was obtained in all, and 57 of the 63 (94%) developed myotubes (Witkowski and Dubowitz, 1975). Improvement in techniques (Witkowski et al., 1976) has contributed to the greater consistency of obtaining positive cultures. We have also been able to slowly freeze down part of our initial biopsy sample for storage in liquid nitrogen for subsequent thawing and successful culture. However, these additional studies have still not shown up any apparent abnormality in diseased muscle in culture. At electron microscopic level we have been unable to detect any abnormalities, even in some of the congenital myopathies such as central core disease and nemaline (rod body) myopathy.

There are a number of possible explanations for the apparent normality of dystrophic and other diseased muscle in culture. One possibility is that the myoblasts which grow and differentiate from explants of diseased muscle are derived from relatively normal fibres within the muscle and that the abnormal fibres do not grow. This seems improbable since in Duchenne muscular dystrophy all fibres are potentially abnormal and will tend to degenerate in the later stages of the disease. Moreover, biopsies taken from older children gave similar results to those from early cases. In one of our cases of central core disease 99% of the fibres in the biopsy showed the presence of cores.

Another possibility is that the myoblasts are in fact abnormal, but that the parameters we have used are not sensitive enough to detect the abnormality. We may be 'barking up the wrong tree'.

A third possibility is that the period in subculture has not been long enough to allow the dystrophic changes to develop. This is perhaps also reflected by the absence of spontaneous contractions in any of these preparations in contrast to the greater maturity and frequent contractions of embryonic animal muscle. Possibly if one could maintain an individual subculture for a much more prolonged period, the myotubes might mature further and eventually manifest dystrophic features.

One of the central problems in human muscle culture is the difficulty of cloning muscle and getting pure muscle devoid of fibroblasts or other tissue. Askanas and Engel (1975) have recently described a 're-explantation' technique which they claim reduces the proportion of fibroblasts derived from the explants.

Finally, there is the interesting possibility that the diseased muscle may in fact revert to a normal state when cultured *in vitro*. This would imply that some environmental factor *in vivo*, such as a neural or humoral influence, may be responsible for perpetuating the dystrophic state of the muscle.

NERVE–MUSCLE PREPARATIONS

The absence of morphological change in dystrophic muscle in culture may simply reflect the relative immaturity of the myotubes in the preparation, since muscular dystrophy normally develops in mature muscles and shows progressive change with time. The recent advent of a new technique described by Peterson and Crain (1972) for innervating muscle in culture, with the establishment of neuromuscular junctions, has opened up the way for a more extensive maturation of muscle in culture. Following innervation these myotubes become more strikingly cross-striated, show peripheral migration of their nuclei and frequently demonstrate spontaneous contraction in culture, in contrast to the absence of contractions in isolated myotubes in culture.

A complete transverse segment of 12–14 day embryonic mouse spinal cord one ganglion thick, with dorsal root ganglia and meninges intact, is first explanted. This is followed by an explant of muscle three or four days later, placed in close proximity to the spinal cord.

Mouse dystrophy

This nerve–muscle model has found application in the study of combined culture of cord and muscle from normal and dystrophic mouse (Gallup and Dubowitz, 1973, 1975; Hamburgh *et al.*, 1973, 1975; Paul and Powell, 1974; see Table 5.1). It was hoped that the various permutations of cord and muscle from normal and dystrophic animals would resolve the question of the relative roles of the muscle itself and the nervous system in the pathogenesis of muscular dystrophy. There are two mutant genes for muscular dystrophy in the mouse, a more severe one 129/ReJdy (dy) and a milder one C57BL/6J dy^{2J} (dy^{2J}). It is probably of some importance in the context of these experiments

Table 5.1 Permutations of normal and dystrophic mouse nerve/muscle cultures

Successful development or function of the muscle in the indicated permutation of spinal cord and muscle is indicated by +, a failure by −. The role of the mutant dy^{2J} cord in muscular dystrophy is thus not confirmed.

	Muscle			
Cord	Normal	dy mutant	dy^{2J} mutant	Authors
Normal	+	+	+	Gallup and Dubowitz (1973, 1975)
	+		+	Hamburgh et al. (1973, 1975)
	+	−		Paul and Powell (1974)
dy^{2J} mutant with mild muscular dystrophy	−	−	−	Gallup and Dubowitz (1973, 1975)
	+		+	Hamburgh et al. (1973, 1975)
dy mutant with severe muscular dystrophy	+	−		Paul and Powell (1974)

that Parsons (1974) found that dy myoblasts would not fuse in culture, whereas dy^{2J} did, suggesting a fundamental difference between these two forms of mouse dystrophy.

These studies have produced some interesting but conflicting results. In our study (Gallup and Dubowitz, 1973, 1975) we found that the normal cord was able to innervate and maintain innervation of either normal or dystrophic muscle (dy and dy^{2J}) and that the degree of maturation of the muscle and the presence of cross-striations and spontaneous contractions was similar. In contrast, cord from dystrophic animals (dy^{2J}), while also showing initial outgrowth of neurites, did not show the same long-term maintenance of innervation or the presence of striations or spontaneous contractions. This suggested that in mouse dystrophy the primary problem might be in the nervous system rather than in the muscle.

However, conflicting results have been obtained by Hamburgh et al. (1973, 1975), who showed that either normal or dystrophic cord could innervate either normal or dystrophic muscle (dy^{2J}), and by Paul and Powell (1974) who found that normal muscle would regenerate in the presence of either normal or dystrophic (dy) spinal cord, but that dystrophic (dy) muscle would not regenerate in the presence of either spinal cord, thus suggesting that the primary abnormality is within the muscle. This is, however, a simplification of their actual results since there was evidence that dy myoblasts did fuse in culture but the myotubes produced were abnormal. There was no difference between dystrophic muscle with normal or dystrophic cord, but the degree of differentiation of these abnormal myotubes seemed to depend on the cord present. Cross-striations were present in some dy myotubes with normal cord but none were found in cultures with dystrophic cord. Motor endplates were not found in these cultures, even after five weeks.

Human nerve/muscle studies

Perhaps of more interest in the context of this symposium is the potential for further study of human muscle provided by the nerve/muscle model. Since there are no immunological barriers in the tissue culture preparation, there is no reason why one should not attempt to culture human muscle in association with animal neurones (Figure 5.1). This has in fact been done and will produce functional neuromuscular junctions which can be identified by cholinesterase stains and also by scanning and transmission electron microscopy.

This provides a means of promoting longer maintenance of the human myotube in culture and one does observe more marked striations, more myofibrils and a migration of the central nuclei to a peripheral position within the myotubes (Witkowski and Dubowitz, 1975). We have also had one culture of normal muscle showing persistent spontaneous contractions, and Peterson and her colleagues (1975, personal communication) have observed this on a number of occasions and have indeed had some preparations continuing to contract rhythmically for up to a year or longer if they were left undisturbed with repeated replenishment of culture medium.

Once again, all our studies of innervated human muscle failed to reveal any pathological change in culture obtained from dystrophic or other diseased muscle (Witkowski and Dubowitz, 1975).

We have recently attempted to culture human neurons with human muscle (Figure 5.2) in the hope of eventually studying directly a combined culture of nerves from a dystrophic fetus in association with normal and dystrophic muscle, in order to throw further light on the potential role of the nervous system in the pathogenesis of human muscular dystrophy.

We have found that the technique used on embryonic mouse spinal cord does not work with human fetal spinal cord, possibly because of the larger size of the cord. By using smaller segments of cord cross-sections we have recently had successful outgrowth for human fetal cord up to 18 weeks' gestation. We have also obtained similar outgrowth from explants of fetal brain. In addition we have been able to freeze nervous tissue, along similar lines used for storage of muscle, and to subsequently thaw it out for successful culture (Statham and Dubowitz, unpublished observations).

It has also been possible to culture dissociated nerve cells, prepared from embryonic brain or spinal cord, in association with muscle with resultant neuromuscular contacts.

Oh (1975, 1976) has recently described the promotion of growth and

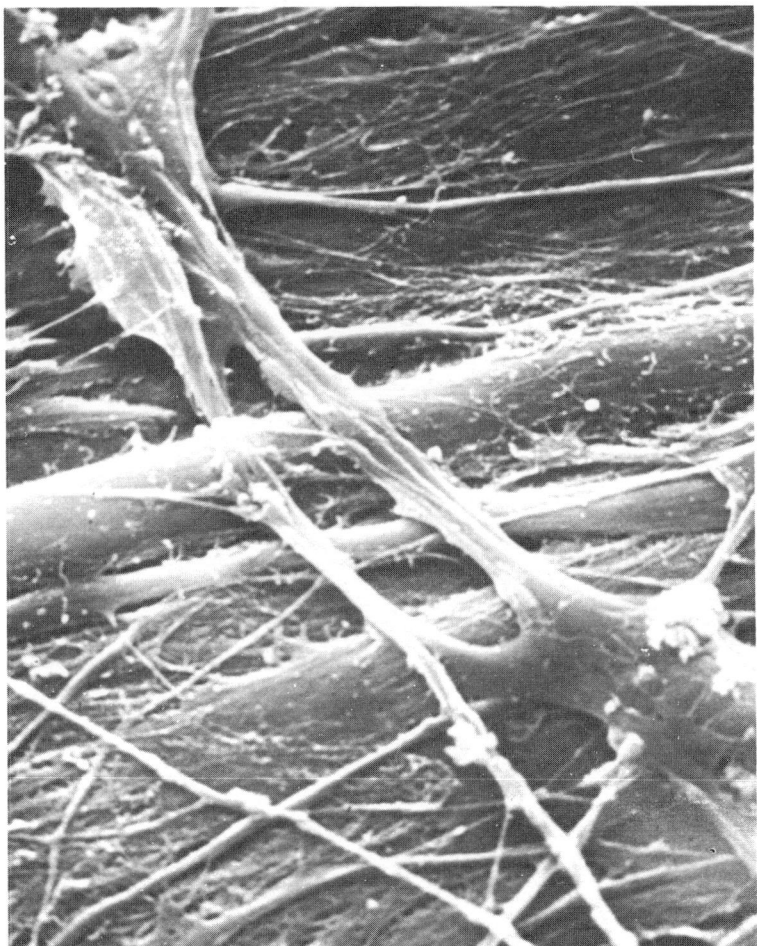

Figure 5.1 Neurites from explant of mouse spinal cord growing in association with human muscle from a muscle biopsy. (Scanning EM, × 2400)

differentiation of chick myoblasts by extracts of chick embryo brain or adult chick sciatic nerve. This may have potential for the growth of human myoblasts, but no comparable studies have yet been made on human muscle.

METABOLIC MYOPATHIES

There are a number of metabolic myopathies in which a specific abnormality has been delineated. The majority are in relation to glycogen metabolism, and muscle involvement may occur in the glycogen storage diseases types II, III, IV, V, VII and VIII (for review see Dubowitz and Brooke, 1973). Myopathies have also

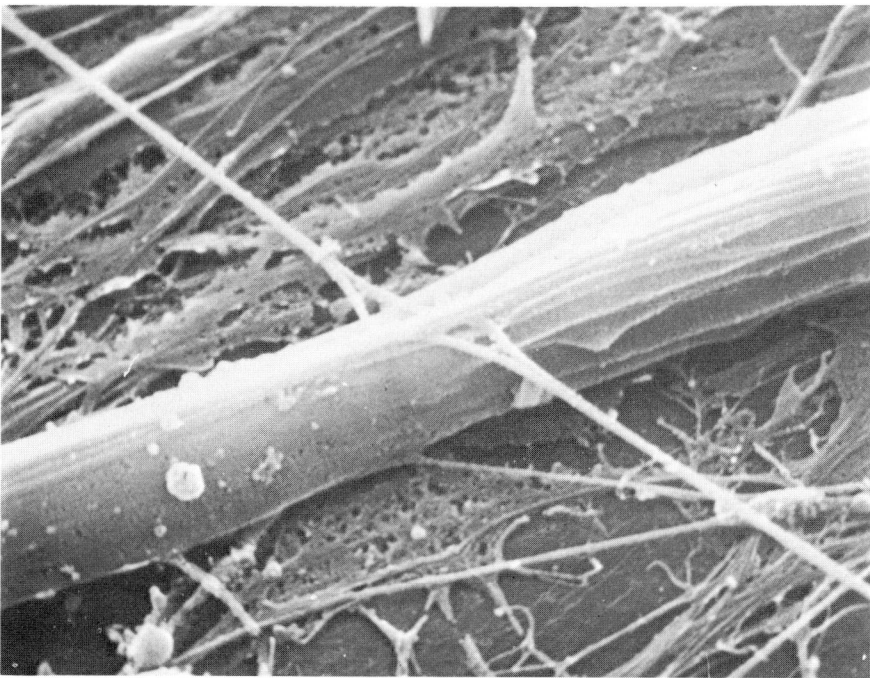

Figure 5.2 Human fetal nerves (from fetal spinal cord) growing in association with human fetal muscle. (Scanning EM, × 8000)

been recognized in association with abnormalities of lipid storage, due to carnitine deficiency and carnitine palmityltransferase deficiency. The large number of clinical muscle syndromes with associated mitochondrial abnormalities probably also have a metabolic basis, although only one or two suggestions have so far been forthcoming in relation to a possible enzyme defect.

A number of these metabolic myopathies have already been studied in culture with some interesting results. Roelofs, Engel and Chauvin (1972) cultured muscle cells from three cases of phosphorylase deficiency (glycogenosis type V, McArdle's disease), and surprisingly found a positive histochemical reaction for phosphorylase in some of the myoblasts in culture, comparable to that of normal muscle, in spite of the complete absence of phosphorylase in the muscle fibres of the biopsy sample. This raised the possibility as to whether 'satellite' cells might be present, which might potentially have phosphorylase present in spite of the absence of phosphorylase in the mature muscle fibre.

In a case of acid maltase deficiency (type II glycogenosis) Askanas and colleagues (1976) found a similar absence of the enzyme in cultured cells to that in the muscle biopsy itself.

CURRENT SHORTCOMINGS: FUTURE POSSIBILITIES

In the human muscular dystrophies it would seem probable that the dystrophic muscle cell should express its phenotype *in vitro*. However, we still have no marker for the dystrophic cell. If one were available it would be of interest to know whether it affects other cells apart from muscle. If fibroblasts were similarly affected this could provide a potential means of identifying the dystrophic fetus *in utero* and a more rational approach to selective termination of affected dystrophic males *in utero* instead of the current termination of all male fetuses in Duchenne carriers.

Application of more sophisticated techniques to the tissue culture model may possibly show up abnormalities not revealed by methods used to date.

Electron microscopy

Electron microscopy of diseased human muscle in culture has to date only had very limited application. It may be of interest to see whether some of the apparent structural abnormalities in the membrane in Duchenne dystrophy biopsies, reported by Mokri and Engel (1975), occur in cultured cells as well.

Application of electron microscopic freeze fracture techniques (Schotland, 1976) have shown differences in membrane granularity in dystrophic muscle, which if confirmed may again have application as a possible marker in the dystrophic cell in culture.

The current vogue is certainly for a membrane abnormality in muscular dystrophy, and the demonstration of decreased phosphorylation in peak III protein of the membrane of the red blood cell in dystrophia myotonica, and an increased phosphorylation of peak II protein in Duchenne dystrophy, has given this theory a recent boost (Roses and Appel 1973; Roses *et al.*, 1975). This awaits confirmation both in the red cell and in other cells. If it is a consistent phenomenon in myotonic dystrophy, and possibly in Duchenne muscular dystrophy, and affects muscle cells and fibroblasts as well, it would have an interesting application to tissue culture.

Recent application of electron microscopic X-ray microprobe analysis to normal and diseased muscle (Yarom *et al.*, 1975; Maunder *et al.*, 1976) has shown potential differences in the subcellular concentration and localization of elements such as calcium between normal and dystrophic muscle and may yield further information when applied to cultures of dystrophic muscle.

Cell behaviour

Observations on the behaviour of human dystrophic cells have been limited to the timing of outgrowth from the explants, appearance of myotubes and cross-striations, and to the length of time cells survive in culture. All these parameters have a wide range which makes it difficult to pick up possible small but significant deviations in individual disorders. Some of the more sophisticated measurements such as accurate measurement of prefusion movements (Powell, 1973) and cell cycle parameters (Yao and Essien, 1975) may be worth applying to human muscle.

Electrophysiological

Detailed studies have been made of the electrophysiological properties of chick muscle (Harris et al., 1973; Ritchie and Fambrough, 1975) and also of the mouse mutant dysgenesis (Powell and Fambrough, 1973). Such studies applied to different human muscle disease in vitro may reflect some fundamental differences in membrane properties.

Biochemistry

Biochemical studies of human muscle in culture have to date been fairly limited and there is scope for further study. Various suggestions have been advanced for a basic abnormality in Duchenne muscular dystrophy, including abnormalities in lipid metabolism (Kunze and Olthoff, 1970, 1972; Hughes, 1972), fructose metabolism (Strickland and Ellis, 1973a,b,c), protein turnover (Kitchin and Watts, 1973) and adenylate cyclase activity (Mawatan et al., 1974), and if any of these can be substantiated in muscle biopsy samples it would be of interest to try to identify a parallel abnormality in the cultured cell.

The myoblast in culture has undoubted potential for further studies of normal and diseased muscle. Perhaps by the time this Society meets for a second symposium on the cell in culture the myoblast will have a more extensive part in the programme than its present rather modest contribution.

ACKNOWLEDGMENTS

This work has been generously supported over a number of years by the Muscular Dystrophy Group of Great Britain. I wish to thank Dr Jan Witkowski, currently at the Mayo Clinic, for his help and advice in the preparation of this manuscript.

References

Askanas, V. and Engel, W. K. (1975). A new program for investigating adult human skeletal muscle grown aneurally in tissue culture. Neurology (Minneap.), 25, 58

Askanas, V., Engel, W. K., Di Mauro, S., Brooks, B. R. and Mehler, M. (1976). Adult-onset acid maltase deficiency. Morphologic and biochemical abnormalities reproduced in cultured muscle. N. Engl. J. Med., 294, 573

Bishop, A., Gallup, B., Skeate, Y. and Dubowitz, V. (1971). Morphological studies on normal and diseased human muscle in culture. J. Neurol. Sci., 15, 183

Dubowitz, V. and Brooke, M. H. (1973). Muscle Biopsy: A Modern Approach. (London and

Philadelphia: Saunders)

Gallup, B., Strugalska-Cynowska, H. and Dubowitz, V. (1972a). Histochemical studies on normal and diseased human and chick muscle in tissue culture. *J. Neurol. Sci.*, **17**, 109

Gallup, B., Bishop, A. and Dubowitz, V. (1972b). Autoradiographic studies of RNA and DNA synthesis during myogenesis in cultures of human, chick and rat muscle. *J. Neurol. Sci.*, **17**, 127

Gallup, B. and Dubowitz, V. (1973). Failure of 'dystrophic' neurones to support functional regeneration of normal or dystrophic muscle in culture. *Nature (London)*, **243**, 287

Gallup, B. and Dubowitz, V. (1975). Regeneration and innervation of normal and dystrophic muscle cultured with normal and dystrophic spinal cord. *Neuropathol. Appl. Neurobiol.*, **1**, 205

Geiger, R. S. and Garvin, J. S. (1957). Pattern of regeneration of muscle from progressive muscular dystrophy patients cultivated *in vitro* as compared to normal human skeletal muscle. *J. Neuropathol. Exp. Neurol.*, **16**, 532

Goyle, S., Kalra, S. L. and Singh, B. (1967). The growth of normal and dystrophic human skeletal muscle in tissue culture. *Neurol. India*, **15**, 149

Hamburgh, M., Bornstein, M. B., Peterson, E. R., Crain, S. M., Masurovsky, E. B. and Kirk, C. (1973). *In vitro* studies of regeneration and innervation of muscle from dystrophic (dy²J) mutant mice. *In*: D. H. Ford (ed.), *Neurobiological Aspects of Maturation and Aging* (Progress in Brain Research, Vol. 40), pp. 497–508 (Amsterdam: Elsevier)

Hamburgh, M., Peterson, E., Bornstein, M. B. and Kirk, C. (1975). Capacity of foetal spinal cord obtained from dystrophic mice (dy²J) to promote muscle regeneration. *Nature (London)*, **256**, 219

Harris, J. B., Marshall, M. W. and Wilson, P. (1973). A physiological study of chick myotubes grown in tissue culture. *J. Physiol (London)*, **229**, 751

Herrmann, H. (1958). An evaluation of tissue culture of muscle. *Res. Publ. Assoc. Res. Nerv. Ment. Dis.*, **38**, 697

Herrmann, H., Konigsberg, U. R. and Robinson, G. (1960). Observations on culture *in vitro* of normal and dystrophic muscle tissue. *Proc. Soc. Exp. Biol. Med.*, **105**, 217

Holtzer, J., Sanger, J. W., Ishikawa, H. and Strahs, K. (1973). Selected topics in skeletal myogenesis. *Cold Spring Harbor Symp. Quant. Biol.*, **37**, 549

Hughes, B. P. (1972). Lipid changes in Duchenne muscular dystrophy. *J. Neurol. Neurosurg. Psychiat.*, **35**, 658

Kakulas, B. A., Papadimitriou, J. M., Knight, J. O. and Mastaglia, F. L. (1968). Normal and abnormal human muscle in tissue culture. *Proc. Aust. Assoc. Neurol.*, **5**, 79

Kitchin, S. E. and Watts, D. C. (1973). Comparison of the turnover patterns of total and individual muscle proteins in normal mice and those with hereditary muscular dystrophy. *Biochem. J.*, **136**, 1017

Kunze, D. and Olthoff, D. (1970). Der Lipidgehalt menschlicher skelettmuskulatur bei Primären und Sekundären myopathien. *Clin. Chim. Acta*, **29**, 455

Kunze, D. and Olthoff, D. (1972). Acylation of lysolecithin with long-chain fatty acids by normal and dystrophic human muscle. *Clin. Chim. Acta*, **36**, 564

Lipton, B. H. and Konigsberg, J. R. (1972). A fine structural analysis of the fusion of myogenic cells. *J. Cell Biol.*, **53**, 348

Maunder, C. A., Dubowitz, V. and Yarom, R. (1976). Electron microscopic X-ray analysis as a potential tool in the localization and quantitation of intracellular elements. *Clin. Sci.*, **50**, 14P

Mawatari, S., Takagi, A. and Rowland, L. P. (1974). Adenyl cyclase in normal and pathologic human muscle. *Arch. Neurol.*, **30**, 96

Mokri, B. and Engel, A. G. (1975). Duchenne dystrophy: electron microscopic findings pointing to a basic or early abnormality in the plasma membrane of the muscle fiber. *Neurology (Minneap.)*, **25**, 1111

Oh, T. H. (1975). Neurotrophic effects: characterization of the nerve extract that stimulates muscle development in culture. *Exp. Neurol.*, **46,** 432

Oh, T. H. (1976). Neurotrophic effects of sciatic nerve extracts on muscle development in culture. *Ex. Neurol.*, **50,** 376

O'Steen, W. K. (1962). Growth activity of normal and dystrophic muscle implants in normal and dystrophic hosts. *Lab. Invest.*, **11,** 412

Parsons, R. (1974). Expression of the dystrophia muscularis (dy) recessive gene in mice. *Nature (London)*, **251,** 621

Paul, C. V. and Powell, J. (1974). Organ culture studies of foetal cord and adult muscle from normal and dystrophic mice. *J. Neurol. Sci.*, **21,** 365

Peterson, E. R. and Crain, S. M. (1972). Regeneration and innervation in cultures of adult mammalian skeletal muscle coupled with fetal rodent spinal cord. *Exp. Neurol.*, **36,** 136

Pogogeff, I. A. and Murray, M. R. (1946). Form and behaviour of adult mammalian skeletal muscle *in vitro*. *Anat. Rec.*, **95,** 321

Powell, J. A. (1973). Development of normal and genetically dystrophic mouse muscle in tissue culture. *Exp. Cell Res.*, **80,** 251

Powell, J. A. and Fambrough, D. M. (1973). Electrical properties of normal and dysgenic mouse skeletal muscle in culture. *J. Cell. Physiol.*, **81,** 21

Ritchie, A. K. and Fambrough, D. M. (1975). Electrophysiological properties of the membrane and acetylcholine receptor in developing rat and chick myotubes. *J. Gen. Physiol.*, **66,** 327

Roelofs, R. I., Engel, W. K. and Chauvin, P. B. (1972). Histochemical phosphorylase activity in regenerating muscle fibers from myophosphorylase-deficiency patients. *Science*, **177,** 795

Roses, A. D. and Appel, S. H. (1973). Erythrocyte protein phosphorylation. *J. Biol. Chem.*, **248,** 1408

Roses, A. D., Herbstreith, M. H. and Appel, S. H. (1975). Membrane protein kinase alteration in Duchenne muscular dystrophy. *Nature (London)*, **254,** 350

Schotland, D. (1976). *In*: L. B. Rowland (ed.), *Pathogenesis of the Human Muscular Dystrophies*, Proceedings of 5th International Scientific Conference sponsored by the Muscular Dystrophy Association, Durango, Colorado, June 21–25, 1976. (Amsterdam: Excerpta Medic) (In press)

Skeate, Y., Bishop, A. and Dubowitz, V. (1969). Differentiation of diseased human muscle in culture. *Cell Tiss. Kinet.*, **2,** 307

Strickland, J. M. and Ellis, D. A. (1973a). The uptake and metabolism of several ^{14}C-labelled sugars by muscle from a normal human foetus and from a human foetus affected with muscular dystrophy. *Biochem. Soc. Trans.*, **1,** 735

Strickland, J. M. and Ellis, D. A. (1973b). The conversion of glucose into fructose in muscle from a normal human foetus and a foetus affected with muscular dystrophy. *Biochem. Soc. Trans.*, **1,** 738

Strickland, J. M. and Ellis, D. A. (1973c). The effects of fructose, glucose 6-phosphate and acetate on the metabolism of (^{14}C)glucose by normal and dystrophic human foetal muscle. *Biochem. Soc. Trans.*, **1,** 740

Witkowski, J. A. and Dubowitz, V. (1975). Growth of diseased human muscle in combined cultures with normal mouse embryonic spinal cord. *J. Neurol. Sci.*, **26,** 203

Witkowski, J. A., Durbidge, M. and Dubowitz, V. (1976). Growth of human muscle in tissue culture. *In vitro*, **12,** 98

Yaffe, D. (1971). Developmental changes preceding cell fusion during muscle differentiation *in vitro*. *Exp. Cell Res.*, **66,** 33

Yao, M-L. and Essien, F. B. (1975). DNA synthesis and the cell cycle in cultures of normal and mutant (mdg/mdg) embryonic mouse muscle cells. *Dev. Biol.*, **45,** 166

Yarom, R., Maunder, C., Scripps, M., Hall, T. A. and Dubowitz, V. (1975). A simplified method of specimen preparation for X-ray microanalysis of muscle and blood cells. *Histochemistry*, **45,** 49

6
Culture of Neurons and Glial Cells

E. J. THOMPSON

Department of Neurochemistry, Institute of Neurology, London

The growth of brain cells in tissue culture is not a new procedure. The technique of tissue culture was first used with central nervous system tissue in order to answer a question about normal developmental sequence, i.e. what is the origin of the axon? (Harrison, 1907).

However, in the context of genetic metabolic alterations in human development, what can central nervous system (CNS) cultures offer? This may be further discussed by posing three specific questions: (1) What is the current state of differentiated function in CNS cultures? (2) What are some of the major limiting factors which must be solved to allow further progress? and (3) What are the particular advantages which CNS cultures offer in exchange for being notoriously more fastidious in their requirements?

Some of the specific functions which have been successfully reproduced in culture are listed, using results obtained from mainly fetal material. This list is based upon cultures initiated from single cells (as opposed to organ cultures, or outgrowth from small tissue pieces). The methods for the preparation of such single cells involve trypsinization or sieving through graded pore sizes. Although it is perhaps more difficult to initiate cultures from single cells, in that it involves a dissociation procedure, most of the functions previously ascribed to organ cultures have now been found to express themselves with cultures derived from single cells, including the generation of complex synaptic activity and multilammelar myelination.

DIFFERENTIATED FUNCTIONS WHICH CULTURED CELLS FROM THE CNS PERFORM

(1) The neurons extend long branching processes which contain neurotubes and microfilaments (Bird and James, 1973).

(2) The neurons generate action potentials (Godfrey *et al.*, 1975).

(3) The neurons respond to γ-aminobutyric acid (GABA), glutamate, glycine and other neurotransmitters when they are iontophoretically applied. These transmitters generate either excitatory postsynaptic potentials or inhibitory postsynaptic potentials (Godfrey et al., 1975).

(4) The neurons synapse with other nerve cells and produce miniature endplate potentials and action potentials. These effects are calcium-sensitive (O'Lague et al., 1974).

(5) Neurons synapse with muscle cells and the electrical activity can be modulated by curare in the expected fashion (Fischback, 1972).

(6) Neuronal endings display characteristic electron microscopic features of the synapse, including synaptic vesicles and postsynaptic thickening (James and Tressman, 1969).

(7) Neurons from sympathetic ganglia bind α-bungarotoxin, a specific marker for acetylcholine receptor protein (Greene et al., 1973).

(8) The neurons synthesize the neurotransmitters, acetylcholine and GABA (Wilson et al., 1972; Patterson and Chun, 1974).

(9) Axons will become myelinated (Bird and James, 1975).

(10) Specificity of cell–cell connections is retained as shown by the studies on specific reaggregation in the hippocampus (Delong, 1970), cerebellum and most recently in nasal and temporal retina which appear to recognize their specific correct left and right tectum (Barbera, 1975).

Various parameters of differentiation have been monitored in order to quantitate the degree to which CNS cultures behave in appropriate functional manner. To the biochemist, obvious indicators of differentiated function are the specific activity of the neurotransmitter enzymes. Enzymatic assay can be more sensitive than assays for specific cell products. In several instances there is sufficient culture material to identify the product of the reaction and thus do proper kinetic analyses. The following enzymes have been studied:

(1) Choline acetyltransferase which synthesizes acetylcholine, the cholinergic transmitter.

(2) Glutamic acid decarboxylase which synthesizes GABA, the inhibitory transmitter.

(3) Acetylcholine esterase which hydrolyzes acetylcholine.

Although on culture one obtains a tenfold increase in the number of cells, protein and DNA (Figure 6.1A) the actual values obtained for the neurotransmitter enzymes have shown successively higher specific activities with further optimization of conditions of cell growth and differentiation. Some earlier results showed the increasing specific activities with further time and differentiation in culture (Figure 6.2). As these cultures matured some cells developed positive staining for Bodian's silver. These cells have very long processes and grow on top of the carpet of other cells (Figure 6.1E). More recent values obtained for specific activities of the neurotransmitter enzymes are higher and approach adult brain values (Table 6.1).

Table 6.1 Specific activities of neurotransmitter enzymes from fetal mammalian brain

Source	Enzymes pMoles product per min per mg protein			Reference
	Choline acetyltransferase	Glutamic acid decarboxylase	Acetylcholine esterase	
Rat	200	Not done	20,000	Godfrey et al., 1975
Rat	Not done	800	11,000	Schrier and Shapiro, 1973

Those cultures which have the highest specific activities for these enzymes also have the highest incidence of electrical activity (Godfrey *et al.*, 1975)

SOME FACTORS WHICH ARE CURRENTLY LIMITING THE FURTHER DIFFERENTIATION OF BRAIN CELLS IN CULTURE

(1) The main limiting factor is that adult tissues (i.e. older than newborn) are much more difficult to propagate than fetal tissues. Further, if one tries to use a more quantitative approach starting with single cells, rather than organ cultures, the actual procedure for dissociation of the tissue is more difficult with adult material. Thus, one must use either collagenase, pronase or papain rather than simple trypsin, the last being quite satisfactory for fetal tissues. Figure 6.3 shows data for cell yield per gram of brain as a function of increasing age of the animal. It can be seen that there is a dramatic reduction in the number of viable cells obtained with increasing age after birth (Wilson *et al.*, unpublished data).

(2) Since individual brain cells have dramatically different functions there is an urgent need for cell specific markers. Further, it would be advantageous to have single cell assays for each cell type. Morphology has been used as a general indicator of function, but this can have notable limitations:

(a) There are 'intermediate' cell types which do not clearly fall into a given category of cell, e.g. between an astrocytic type and a neuronal type.

(b) Cell shape is not a reliable indication of a cell's potential for assuming a more differentiated form, e.g. ostensibly dedifferentiated cells (also called 'fibroblasts') may fully differentiate with extensive process formation after appropriate manipulation of the culture conditions (Shapiro, 1973).

(c) One type of cell may change into another type, e.g. oligodendrocytes into astrocytes (Nakai, 1963).

(d) Another remaining difficulty is how to objectively quantitate the degree of differentiation for any given cell.

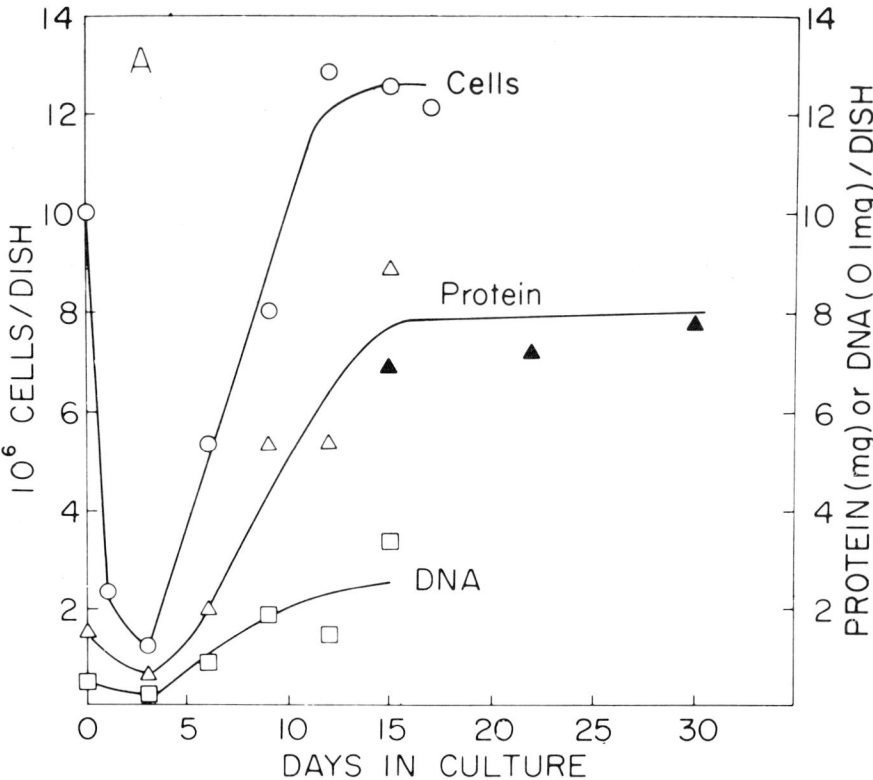

Figure 6.1.A.   Growth characteristics and morphology of cultured mouse brain cells. Mixed cells from brains of newborn mice were obtained and cultured as described (Wilson *et al.*, 1972). *A*, (**O**) cell number (determined by counting trypsinized cells); (**△, ▲**) protein and (**□**) DNA contents of dishes at various times in culture. Values at zero time are those determined for the inocula. *Open symbols,* average of determinations on eight culture dishes (four from each of two separate, simultaneous time curves); *closed symbols,* average of determinations on four culture dishes from a third time curve

For example, theta antigen which increases with myelination *in vivo* has been found to be generally distributed on several brain cell types (Mirsky and Thompson, 1975). The problem probably lies with finding cell specific proteins like the 'acetylcholine receptor' which can then be used as an immunogen (Thompson, 1976).

(3) Further work is required on the standardization of the type, quantity and time of use of culture additives. For example should fetal calf serum or horse serum with or without heat inactivation be used and in what concentration and when should additions be made? Although chick embryo extract can have widely variable effects on cultures, we have recently found that heat inactivation and Amicon filtration (to collect proteins larger than 10 000 daltons) gives

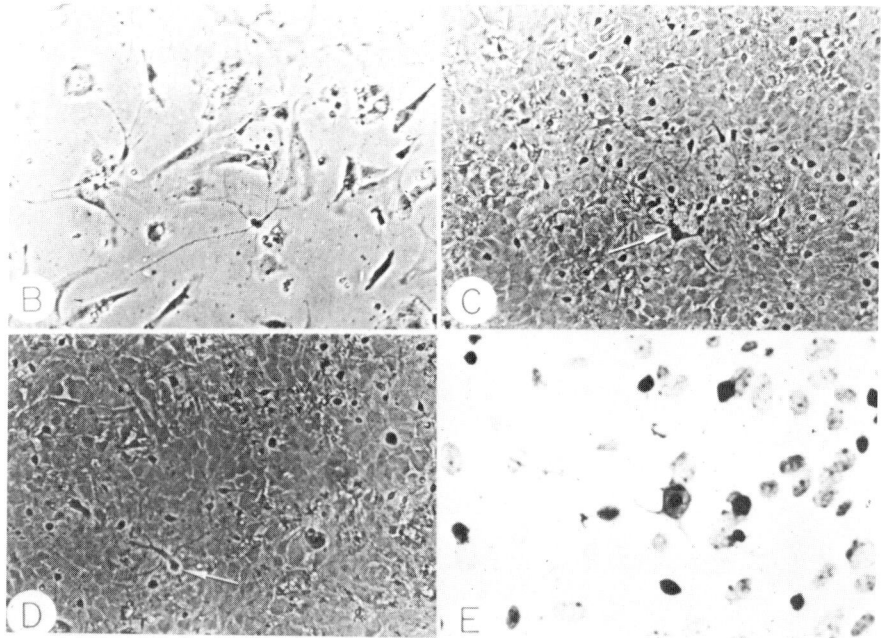

Figure 6.1.B–E.   *B* to *D*, Phase-contrast photomicrographs of living cultures from the time curves. Magnifications of the 35-mm negatives were × 50. *B*, 3 days in culture. *C*, 11 days in culture, *arrow* indicates phase-dark cell (see text). *D*, 11 days in culture, *arrow* indicates phase-bright cell (see text). *E*, Bodian stain (modified protargol) of similar cells grown on cover glasses for 42 days. Bright field microscopy; magnification of the 4 × 5 inch negative was × 600

a preparation with quite reproducible culture results (Skrbic *et al.*, 1975). The support on which the cell grows is an important variable; collagen or gelatin has been used as well as histone, polylysine and polyornithine. The use of these polymers suggests that a basic charge is important for the cell growth surface (Letourneau, 1975).

WHAT WOULD CNS CULTURES OFFER TO THE STUDY OF INHERITED METABOLIC DISEASES?

In the case of the lipidoses, one is usually concerned with catabolic enzyme deficiencies, that is, missing lysosomal hydrolases. Polymorphonuclear leucocytes are often used for assay of the specific enzyme (Harkness, p. 90) but 'T' lymphocytes may be better because these cells are genetically programmed to synthesize at least some of the more complex gangliosides (Esselman and Miller, 1974).

However, a most exciting new chapter in the history of the inborn errors of metabolism has been opened with the finding of an anabolic deficiency, i.e. the storage of G_{M3} ganglioside because of the missing microsomal enzyme which

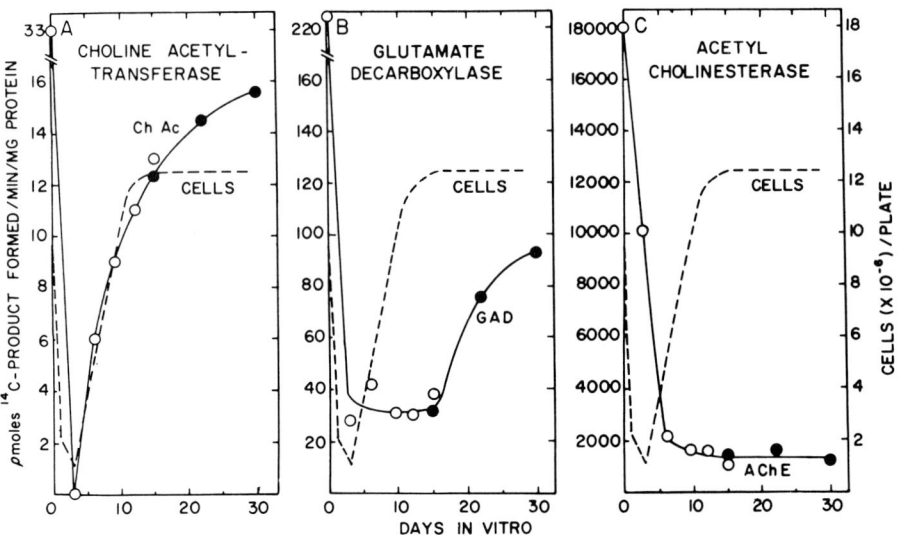

Figure 6.2 Development of marker enzyme specific activities in surface cultures of newborn mouse brain cells. Each homogenate of Figure 6.1.A. was evaluated for the marker enzyme activities as described (Wilson *et al.*, 1972). Changes in specific activities (O , ●) and cell number (*dashed lines*) with time are shown for: *A*, glutamate decarboxylase (assessed by $^{14}CO_2$ evolution by Method *b*); *B*, choline acetyltransferase; and *C*, acetylcholinesterase. One unit of activity is defined as 1 pmole of product formed per min per mg of homogenate protein. *Open* and *closed symbols* are explained in the legend to Figure 6.1.A.

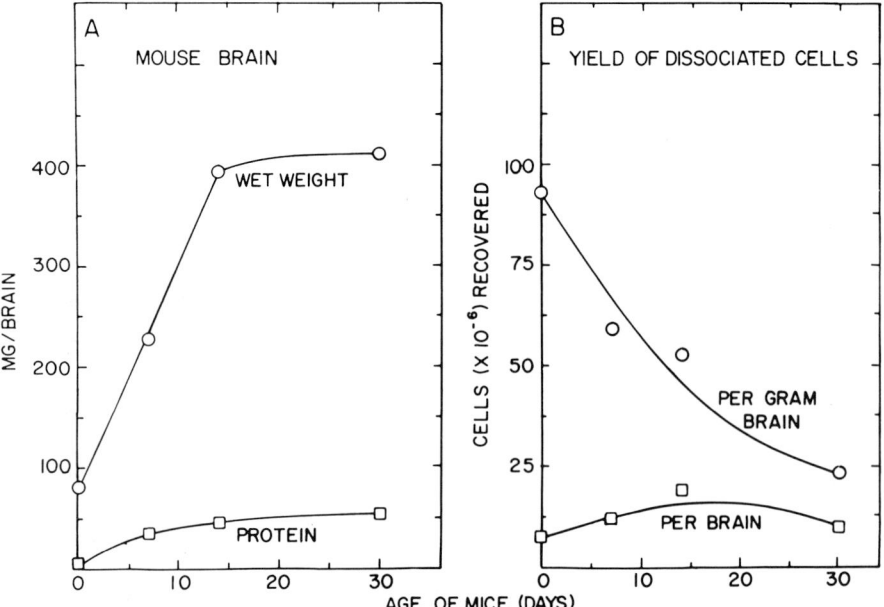

Figure 6.3 (A); Increasing weight of mouse brain with age. (B); Decreasing yield of dissociated cells per gram of brain with age

further builds on to the G_{M3} substrate, rather than the loss of a lysosomal enzyme which degrades this specific product (Fishman *et al.*, 1975). Since this disease (which might have been histologically termed Canavan's spongy degeneration) also affects the glial cells (Tanaka *et al.*, 1975), one useful application of a differentiated CNS culture, which is actively synthesizing these substrates, would be to serve as a source of the appropriate microsomal enzymes.

In summary, although there is clearly scope for further work on expression of differentiated function in CNS cultures, such material will hopefully play an increasing diagnostic role, particularly in anabolic defects of metabolism.

References

Barbera, A. J. (1975). Adhesive recognition between developing retinal cells and optic tecta of chick embryos. *Dev. Biol.*, **46**, 167

Bird, M. M. and James, D. W. (1973). The development of synapses *in vitro* between previously dissociated chick spinal cord neurons. *Z. Zellforsch.*, **140**, 203

Bird, M. M. and James, D. W. (1975). Myelin formation in cultures of previously dissociated mouse spinal cord. *Cell Tiss. Res.*, **162**, 93

Delong, G. R. (1970). Histogenesis of fetal mouse isocortex and hippocampus in reaggregating cell cultures. *Dev. Biol.*, **22**, 563

Esselman, W. J. and Miller, H. C. (1974). Brain and thymus lipid inhibition of antibrain-associated theta cytoxicity. *J. Exp. Med.*, **139**, 445

Fischbach, G. D. (1972). Synapse formation between dissociated nerve and muscle cells in low density cell cultures. *Dev. Biol.*, **28**, 407

Fishman, P. H., Max, S. R., Tallman, J. F., Brady, R. O., Maclaren N. K. and Cornblath, M. (1975). Deficient ganglioside biosynthesis: a novel human sphingolipidosis. *Science*, **187**, 68

Godfrey, E. W., Nelson, P. G., Schrier, B. K., Breuer, A. C. and Ransom, B. R. (1975). Neurons from fetal rat brain in a new cell culture system: a multidisciplinary analysis. *Brain Res.*, **90**, 1

Greene, L. A., Sytkowski, A. J., Vogel, Z. and Nirenberg, M. W. (1973). α-Bungarotoxin used as a probe for acetylcholine receptors of cultured neurones. *Nature (London)*, **243**, 163

Harrison, R. G. (1907). Observations on the living developing nerve fibre. *Proc. Soc. Exp. Biol. Med.*, **4**, 140

James, D. W. and Tressman, R. L. (1969). An electron microscopic study of the *de novo* formation of neuromuscular junctions in tissue culture. *Z. Zellforsch.*, **100**, 126

Letourneau, P. C. (1975). Cell-to-substratum adhesion and guidance of axonal elongation. *Dev. Biol.*, **44**, 92

Mirsky, R. and Thompson, E. J. (1975). Thy 1 (Theta) antigen on the surface of morphologically distinct brain cell types. *Cell*, **4**, 95

Nakai, J. (ed.). (1963). *Morphology of Neuroglia*, Chapter III, pp. 65–74. Glia Research Group, sponsored by Ministry of Education, Japan (Tokyo: Igaku Shoin Ltd.)

O'Lague, P. H., Obata, K., Claude, P., Furshpan, E. J. and Potter, D. D. (1974). Evidence for cholinergic synapses between dissociated rat sympathetic neurons in cell culture. *Proc. Nat. Acad. Sci. U.S.A.*, **71**, 3602

Patterson, P. H. and Chun, L. L. Y. (1974). The influence of non-neuronal cells on catecholamine and acetylcholine synthesis and accumulation in cultures of dissociated sympathetic neurons. *Proc. Nat. Acad. Sci. U.S.A.*, **71**, 3607

Schrier, B. K. and Shapiro, D. L. (1973). Effects of N^6-monobutyryl-cyclic AMP on glutamate decarboxylase activity in fetal rat brain cells and glial tumour cells in culture. *Exp. Cell Res.*, **80**, 459

Shapiro, D. L. (1973). Morphological and biochemical alterations in foetal rat brain cells cultured in the presence of monobutyryl cyclic AMP. *Nature (London)*, **241**, 203

Skrbic, T. R., Yasin, R., van Beers, G., Bulien, D. and Thompson, E. J. (1975). Optimization of tissue-culture conditions for differentiation of muscle from dissociated single cells. *Biochem. Soc. Trans.*, **3,** 496

Tanaka, J., Garcia, J. H., Max, S. R., Viloria, J. E., Kamijyo, Y., McLaren, N. K., Cornblath, M. and Brady, R. O. (1975). Cerebral sponginess and GM_3 gangliosidosis: ultrastructure and probable pathogenesis. *J. Neuropathol. Exp. Neurol.*, **34,** 249

Thompson, E. J. (1976). Brain specific antigens: biochemical role in selective pathogenesis. *In*: A. N. Davison (ed.). *Biochemistry and Neurological Disease*, pp. 278–316. (Oxford: Blackwell Scientific Publications)

Wilson, S. H., Schrier, B. K., Farber, J. L., Thompson, E. J., Rosenberg, R. N., Blume, A. J. and Nirenberg, M. W. (1972). Markers for gene expression in cultured cells from the nervous system. *J. Biol. Chem.*, **247,** 3159

Biochemistry

7
The Biochemistry of Cytodifferentiation: an Outline of Progress

D. E. S. TRUMAN

Department of Animal Genetics, University of Edinburgh

In this review I would like to outline present knowledge of the regulatory processes in cytodifferentiation, to give a personal view of what might be the most immediate advances in our knowledge of the phenomenon, to suggest what might be of especial interest to those concerned with inborn errors of metabolism, and, finally, to point at least one moral for the experimentalist investigating inherited differences in metabolism.

We can review the process of cell differentiation by examining the flow of information that must be involved as the metabolism of a cell is directed by its genes and as the genes themselves are regulated. Figure 7.1 places the genes in a central position and shows the flow of information from the genes, via messenger RNA (mRNA) to the proteins which regulate cell metabolism and behaviour. This pathway is now familiar in outline and more detail is continually being added to our knowledge of the machinery which translates the message of the mRNA into protein. There are numerous possibilities of special regulatory mechanisms at this level, affecting ribosomes, the state of the mRNA and the rate of its breakdown. While experiments with the injection of exogenous mRNA indicate that the rate of specific protein synthesis may be related to the quantity of mRNA (Gurdon *et al.*, 1971), and other experiments show that specific response to hormones may be a direct function of the quantity of new mRNA transcribed (O'Malley *et al.*, 1975), there is some evidence that the efficiency of translation of messages may vary with different types of mRNA and may change during development (Clayton *et al.*, 1974).

The isolation and characterization of mRNA continues in many laboratories

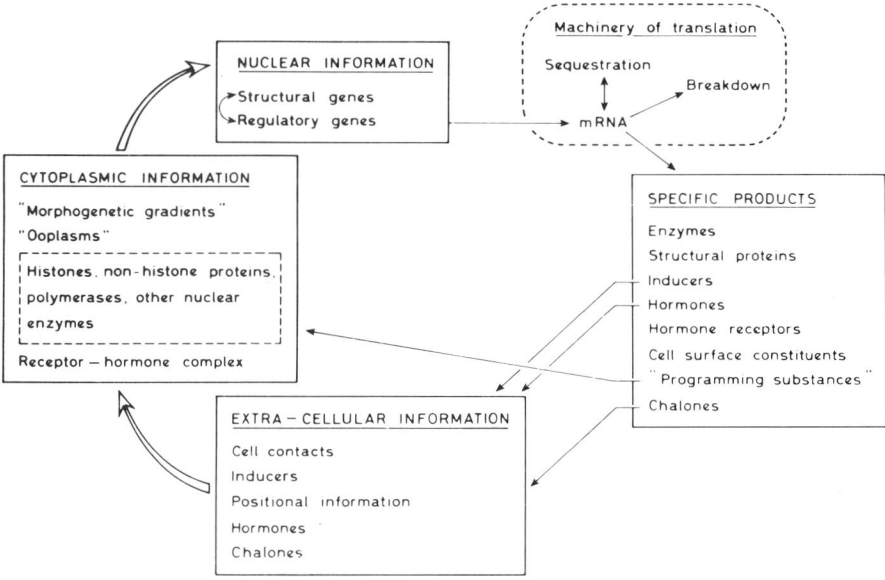

Figure 7.1   Possible pathways of information flow during differentiation

and recent years have seen the application of new techniques in the preparation, analysis and assay of mRNA. The use of reverse transcriptase to make strands complementary to isolated mRNA (i.e. cDNA) has permitted accurate measurement of small amounts of mRNA. The technique has been much used with globin mRNA (Harrison *et al.*, 1974) and has been central to several studies which have elucidated the nature of the lesions of a number of different forms of thalassaema. For example, while it has been found that some cases of α-thalassaemia are associated with the absence of α globin genes, in some cases of β-thalassaemia the β globin genes may be present while the mRNA for β globin is undetectable (Tolstoshev *et al.*, 1976).

Of even greater potential in studies of mRNA are the techniques of genetic engineering whereby preparations of mRNA can be increased in amount almost indefinitely by incorporation, indirectly, into the genome of bacteria. Again globin mRNA has been favoured for such study (Rabbitts, 1976). Experiments of this type must be carried out with great caution, but their potential value is great. Similar techniques can also be used for the building up of quantities of genes from nucleated cells, eukaryotes, by causing them to replicate within microorganisms without well defined nuclei, prokaryotes (Kedes *et al.*, 1975).

NUCLEAR INFORMATION

Let us now turn to the genetic material of the mammalian cell. The relatively independent genetic and translation systems of mitochondria have received the attention of their adherents for a long time, but I shall disregard them for our present purpose and consider the nuclear genes which currently are the focus of those concerned with differentiation. Theoretical models such as those of Britten and Davidson (1969, 1971) continue to provide some inspiration and they have been gradually clothed with more molecular detail (e.g. Paul, 1972), though at present such schemes seem capable neither of proof nor disproof. Molecular hybridization studies of DNA have already yielded a great deal of information about the frequency of different types of sequences in the mammalian genome. The middle repetitive sequences, present in many copies but with some sequence complexity, seem likely candidates for a regulatory role. Even the single copy DNA may not all be used in coding for protein and so may be involved in control (Rosbash et al., 1975).

Perhaps the best evidence that we could have for the support of a genetic model involving regulatory genes directing groups of structural genes would be the discovery of a mutant which was clearly not of a structural gene but was of a regulator. It seems probable that it would be clinicians who might first direct us to such a mutant. What we might expect would be that a group of functionally related enzymes, such as those of the urea cycle, would collectively show some aberrant behaviour. Such a group of enzymes might be absent from a tissue where they normally occur, or more convincingly, all be present in a tissue from which they are normally absent. The presence or absence of structural genes could, in principle, be confirmed by techniques now available, although at present it is only structural proteins and not enzymes which have been the subjects of successful molecular studies of mRNA and genes.

The molecular mechanism by which gene activity is repressed or activated continues to receive attention. The interaction of DNA, histone and nonhistone proteins of the chromatin appears to be involved in the mechanism which determines whether or not a particular portion of DNA is transcribed. The earlier views that the histones had a repressive function, albeit with their specificity determined by nonhistone protein, must now apparently be rejected since inactive and active chromatin seem to have very similar structure, with no apparent difference in histone content (McCarthy et al., 1974). The hypothesis concerning the influence of nonhistone proteins on template specificity of isolated chromatin (MacGillivray et al., 1972), has received the support of numerous studies of the complexity of the nonhistone proteins (Stein et al., 1974). Further evidence has been provided by the variations found in the nonhistone proteins between different tissues, during the cell cycle, and in response to hormone treatment (Wang and Nyberg, 1974).

SPECIFIC PRODUCTS

The role of enzymes and of specific structural proteins, such as collagen, actin or crystallin, in tissue specificity is well known (Figure 7.1). However, I would like to stress that the regulation of the quantities of specific proteins, as opposed to their presence, may be just as significant in determining cell metabolism or behaviour (Truman, 1974). In Figure 7.1 other specific products are mentioned which may be less immediately obvious. Some of the substances conveying signals during embryonic induction appear to be proteins; for example the 'vegetalizing factor' which induces amphibian gastrula ectoderm to give rise to tissues such as notochord, muscle and renal tubules (Tiedemann, 1971).

Hormones are much better characterized substances and where they are not proteins or peptides we can regard them as the products of enzymatically regulated metabolism. Just as some tissues are specialized to act as secretors of hormones, so others include as part of their characteristics the fact that they are target tissues. This attribute may also depend on the proteins which these tissues synthesize. Some hormone receptors are proteins found within the cytosol, such as the estrogen receptor of the uterus or the progesterone receptor of chick oviduct (O'Malley and Means, 1974) while others, such as the insulin receptor, are bound to the cell membrane (Cutrecasas, 1973).

The nature of the cell surface is receiving a great deal of attention and many different approaches are being tried. Chemical studies of the isolated membrane, electron microscopy and immunological methods are all yielding results, while the very specific binding to carbohydrate groups of certain substances, the lectins, has offered a tool which has proved most successful in cell surface studies (Nicolson, 1974). From studies of lectin binding, it is evident that tissues differ in some of the groups exposed on the surface of their cells, that during development these groups may change, and that malignancy may involve surface changes detectable by this method (Inbar and Sachs, 1969; Moscona, 1974).

The 'programming substances' mentioned in Figure 7.1 will be considered in the section on the stability of differentiation.

Chalones are substances in need of further characterization. They appear to be small proteins that are tissue specific, but not species specific, and that rapidly diffuse away from a small mass of cells. When the cell mass is larger they disperse less readily, their concentration builds up and they then exert an inhibitory action on the rate of cell division, so that they act as a signal in a negative feedback loop that tends to maintain an organ at a certain size (Bullough, 1975).

EXTRACELLULAR INFORMATION

Some extracellular information influencing differentiation is shown in Figure 7.1.

It has long been apparent that the fate of a cell may be influenced by its near neighbours. The classical experiments on embryonic induction imply this and in more recent times the nature of the interaction has been investigated in experiments such as those of Grobstein and his colleagues (Fell and Grobstein,

1968). The effect of neighbouring cells may be separated into the effect of different tissues which may act as inducers, and the effect of a mass of adjacent similar cells. It seems that the inductive effect of one tissue on another may not normally require cell contact, but is probably effected by macromolecules which diffuse over small distances (Grobstein, 1964). The significance of the presence of similar adjacent cells is seen in the requirement for a minimal cell mass to effect a response to induction in the case of nephrogenic mesenchyme in the mouse embryo (Wessels and Cohen, 1967). Another effect of cell mass is seen when lens epithelial cells are grown in culture. Those cells which have most contact with other cells are the first to synthesize specific crystallins (Clayton *et al.*, 1976a). A further example is found in cultured embryonic neural retina cells. Whereas aggregates of these cells respond to the action of hydrocortisone by synthesizing glutamine synthetase, confluent but monolayer cultures of similar cells show no such response (Morris and Moscona, 1971).

There are many experiments which indicate that the fate of cells may depend on their position within a cell mass, and it has been proposed that cells must be in receipt of 'positional information', though the form of this information remains unclear (Wolpert *et al.*, 1972).

CYTOPLASMIC INFORMATION

Whereas the nature of the flow of information from the nucleus to the cytoplasm is known in detail, with yet more detail constantly being added, the nature of the information flow from cytoplasm to nucleus is embarrassingly obscure (Figure 7.2). That some information must pass to the nucleus has been apparent since the experiments of Boveri in 1899, and classical experimental embryology has long shown that the fate of a nucleus may depend upon the cytoplasm which surrounds it. The embryologists have spoken of 'morphogenetic gradients' and 'ooplasms' but the biochemical nature of these remains obscure. Of course certain specific macromolecules must pass from the cytoplasm to the nucleus since such significant proteins as the histones, the nonhistone proteins of the chromatin, DNA- and RNA-polymerases and other nuclear enzymes are synthesized in the cytoplasm.

The most valuable techniques for the study of the effect of the cytoplasm on the nucleus have been nuclear transplantation and cell fusion. The oocyte of *Xenopus laevis* provides an environment in which an injected nucleus can retain activity for a prolonged period. An injected nucleus from a HeLa cell will continue to synthesize RNA and this may be translated in the cytoplasm to produce detectable amounts of human proteins. The nucleus takes up proteins from the host cytoplasm and will even incorporate heterologous protein which has been injected into the oocyte (Gurdon *et al.*, 1976).

It is clear from both transplantation and fusion experiments that the cytoplasm has a marked effect on DNA and RNA synthesis (Gurdon and Woodland, 1968; Harris *et al.*, 1969). An extreme example of the effect of the cytoplasm is seen when the nucleus of an avian erythrocyte, which normally synthesizes neither RNA nor DNA, is brought under the influence of the cytoplasm of a HeLa cell. Under these circumstances the avian nucleus shows an expansion in size, decondensation of its DNA and eventually active DNA and RNA synthesis which can lead to the synthesis of haemoglobin. The changes that occur have been analysed by Appels and co-workers (1974). Histone of the f_1 fraction enters the nucleus, and this has the immunological specificity of human histone, while chick f_2 histone leaves the nucleus. Nonhistone protein also enters the nucleus in some quantity, especially a class known as fraction II protein. However, the pattern of regulation of protein synthesis in fused cells is not a simple one and the synthesis of different proteins may be affected differentially by the cytoplasmic environment. For example, when primitive chick erythroid cells are fused with established cell lines from the mouse or hamster, the synthesis both of haem and of globin declines steadily and eventually ceases while the synthesis of other chick proteins in the heterokaryon continues (Davis and Harris, 1975).

These studies illustrate the detailed work being carried out to seek the nature of the signals that pass from cytoplasm to nucleus. The nucleus accepts many components from the cytoplasm and of these the nonhistone proteins of the chromatin may have special significance. The techniques may now be available for rapid advances.

STABILITY OF DIFFERENTIATION

One of the characteristics of differentiated cells is that if they divide their progeny usually show a similar differentiated state, or they may continue along a pathway that is characteristic of that cell type, as when cultured myoblasts fuse to form myotubes (Konigsberg, 1963; Yaffe, 1969) or when lens epithelial cells give rise to lens fibres in lentoid formation in culture (Okada et al., 1971). Exceptions to this general rule of stability are of great interest and theoretical importance. The differentiation of lens-like cells from the pigmented epithelium of iris (Eguchi et al., 1974), from pigmented cells of the retina (Eguchi and Okada, 1973) and from neural retina (Okada et al., 1975) all provide tools which may be invaluable in molecular studies of differentiation. Nevertheless, they are exceptions and our general picture of the controls in cytodifferentiation must accommodate the notion of its stability. While it would be possible to maintain that this stability could be achieved in the whole organism by the homeostatic effect of the cellular environment, the stability observed in the cloned muscle or lens epithelial cells implies that information must be carried from one cell generation to the next. Hence we might postulate some 'programming substance' produced in a cell which influences the next generation. This idea has been explored to some extent by Gurdon and Woodland (1970), and their model supposed that the programming substances were normally present in the cytoplasm with their access to the nucleus limited to certain phases of

the cell cycle. An alternative hypothesis has concentrated on the tertiary structure of DNA in chromatin as a means of producing an inheritable regulation of
DNA template function (Cook, 1973). However this model seems more
applicable to X chromosome inactivation than to the finer control of
differentiation.

These aspects of differentiation seem to have attracted little attention by
molecular biologists and not only is there no progress to report, but I feel little
optimism that the near future will see us much nearer an understanding of the
stability of the differentiated state.

INTERACTIONS OF GENOTYPE AND CELLULAR ENVIRONMENT

The theme which I would like to emphasize is that of the interaction of many
variables to produce the cellular phenotype. Whereas in some species of in-
vertebrates, such as molluscs, cells may have a relatively determined fate of
development programmed by intrinsic information, in higher animals the interactions between the cell and its environment regulate the fate of the cell and
the course of its differentiation. I would like to end by giving an example of interaction between intrinsic and extrinsic information which may contain a
warning to anyone trying to study the nature of the effect of genetic differences
on cell metabolism. The example is from the lens of the eye of the chick and is a
comparison of lenses from normal birds with lenses from birds showing congenital hyperplasia of the lens epithelium (Clayton, 1975). Data shown in
Figure 7.2 are from a study of nucleic acid metabolism in hyperplastic and normal lenses (Truman et al., 1976) and we see that the rate of incorporation of
uridine into RNA varies according to the time that the lenses have been in the
culture medium. It is also known that overall rates of protein synthesis and the
pattern of proteins synthesized are responsive to the culture conditions
(Clayton et al., 1976b). What I want to emphasize here is that the genotype
affects the response to culture conditions and that if we had simply looked at
the amount of incorporation over a period of time we would have come to
different conclusions about the relative synthetic activity, according to the time
taken for the experiment. Thus, the lowest curve in Figure 7.2 shows that the
relative incorporation in normal and abnormal varies according to the interval
chosen for study and that had we measured this particular parameter with a
$3\frac{1}{2}$ h incubation period we would have seen no difference between the strains of
chicks.

Clearly the study of tissues and cells in culture is an invaluable technique in
trying to elucidate the nature of genetic variations in metabolism, but because
of all the interactions which normally occur in the whole organism the
behaviour of material in culture may show features which would not be found
in vivo. If we are skilful, perceptive and lucky we may be able to gain fresh in-
sights even from these anomalies.

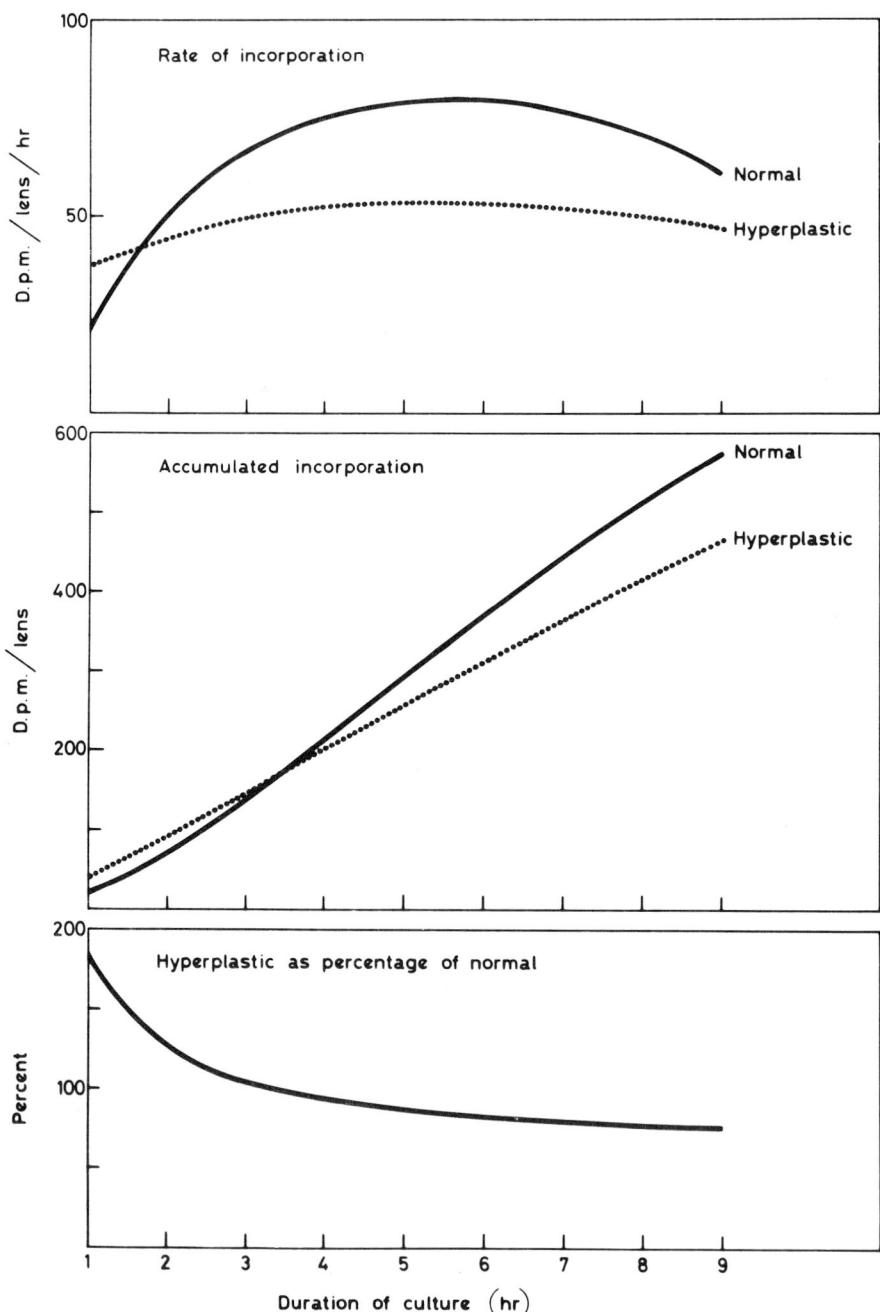

Figure 7.2   Effect of duration of culture on apparent differences in incorporation of uridine in RNA in normal lenses and those with epithelial hyperplasia

ACKNOWLEDGEMENTS

I am grateful to many colleagues for helpful discussions, especially to J. C. Campbell, R. M. Clayton and I. Thomson. The work of the research group on the lens at the Institute of Animal Genetics, Edinburgh, is supported by grants from the Cancer Research Campaign and the Medical Research Council.

References

Appels, R., Bolund, L. and Ringertz, N. R. (1974). Biochemical analysis of reactivated chick erythrocyte nuclei isolated from chick/HeLa heterokaryons. *J. Mol. Biol.*, **87**, 339

Britten, R. J. and Davidson, E. H. (1969). Gene regulation for higher cells: a theory. *Science*, **165**, 349

Britten, R. J. and Davidson, E. H. (1971). Repetitive and non-repetitive DNA sequences and a speculation on the origins of evolutionary novelty. *Qu. Rev. Biol.*, **46**, 111

Bullough, W. S. (1975). Mitotic control in adult mammalian tissue. *Biol. Rev.*, **50**, 99

Clayton, R. M. (1975). Failure of growth regulation of the lens epithelium in strains of fast-growing chicks. *Genet. Res.*, **25**, 79

Clayton, R. M., Eguchi, G., Truman, D. E. S., Perry, M. M., Jacob, J. and Flint, O. P. (1967a). Abnormalities in the differentiation and cellular properties of hyperplastic lens epithelium from strains of chickens selected for high growth rate. *J. Embryol. Exp. Morphol.*, **35**, 1

Clayton, R. M., Odeigah, P. G., de Pomerai, D. I., Pritchard, D. J., Thomson, I. and Truman, D. E. S. (1976b). Experimental modifications of the quantitative pattern of crystallin synthesis in normal and hyperplastic lens epithelia. *In*: Y. Courtois and F. Regnault (eds.). *Biology of the Epithelial Lens Cells in Relation to Development, Ageing and Cataract.* (Paris: Les Colloques de l'I.N.S.E.R.M.) 123

Clayton, R. M., Truman, D. E. S. and Hannah, A. I. (1974). RNA turnover and translational regulation of specific crystallin synthesis. *Cell Differ.*, **3**, 135

Cook, P. R. (1973). Hypothesis on differentiation and the inheritance of gene superstructure. *Nature (London)*, **245**, 23

Cutrecasas, P. (1973). Insulin receptor of liver and fat cell membranes. *Fed. Proc. Fed. Am. Soc. Exp. Biol.*, **32**, 1838

Davis, T. J. and Harris, H. (1975). Haemoglobin synthesis in fused cells. *J. Cell Sci.*, **18**, 207

Eguchi, G., Abe, S. and Watanabe, K. (1974). Differentiation of lens-like structures from newt iris epithelial cells *in vitro*. *Proc. Nat. Acad. Sci. U.S.A.*, **71**, 5052

Eguchi, G. and Okada, T. S. (1973). Differentiation of lens tissue from the progeny of chick retinal pigment cells cultured *in vitro*: a demonstration of a switch of cell types in clonal cell culture. *Proc. Nat. Acad. Sci. U.S.A.*, **70**, 1459

Fell, P. E. and Grobstein, C. (1968). The influence of extra-epithelial factors on the growth of embryonic mouse pancreatic epithelium. *Exp. Cell Res.*, **53**, 301

Grobstein, C. (1964). Cytodifferentiation and its control. *Science*, **143**, 643

Gurdon, J. B., De Robertis, E. M. and Partington, G. (1976). Injected nuclei in frog oocytes provide a living cell system for the study of transcriptional control. *Nature (London)*, **260**, 116

Gurdon, J. B., Lane, C. D., Woodland, H. R. and Marbaix, G. (1971). Use of frog eggs and oocytes for the study of messenger RNA and its translation in living cells. *Nature (London)*, **233**, 177

Gurdon, J. B. and Woodland, H. R. (1968). The cytoplasmic control of nuclear activity in animal development. *Biol. Rev.*, **43**, 233

Gurdon, J. B. and Woodland, H. R. (1970). On the long-term control of nuclear activity during cell differentiation. *Curr. Top. Dev. Biol.*, **5**, 39

Harris, H., Sidebottom, E., Grace, D. M. and Bramwell, M. (1969). The expression of genetic information: a study with hybrid animal cells. *J. Cell Sci.*, **4**, 499

Harrison, P. R., Birnie, G. D., Hell, A., Humphries, S., Young, B. D. and Paul, J. (1974). Kinetic studies of gene frequency. I. Use of a DNA copy of reticulocyte 9S RNA to estimate globin gene dosage in mouse tissues. *J. Mol. Biol.*, **84**, 539

Inbar, M. and Sachs, L. (1969). Interaction of the carbohydrate-binding protein concanavalin A with normal and transformed cells. *Proc. Nat. Acad. Sci. U.S.A.*, **63**, 1418

Kedes, L. H., Chang, A. C. Y., Houseman, D. and Cohen, S. N. (1975). Isolation of histone genes from unfractionated sea urchin DNA by subculture cloning in *E. coli. Nature (London)*, **255**, 533

Konigsberg, I. R. (1963). Clonal analysis of myogenesis. *Science*, **140**, 1273

MacGillivray, A. J., Paul, J. and Threlfall, G. (1972). Transcription regulation in eukaryotic cells. *Adv. Cancer Res.*, **15**, 93

McCarthy, B. J., Nishiura, J. T., Doenecke, D., Nasser, D. S. and Johnson, C. B. (1974). Transcription and chromatin structure. *Cold Spring Harbor Symp. Quant. Biol.*, **38**, 763

Morris, J. E. and Moscona, A. A. (1971). The induction of glutamine synthetase in aggregates of embryonic neural retina cells: correlation with differentiation and multicellular organization. *Dev. Biol.*, **25**, 420

Moscona, A. A. (1974). Surface specification of embryonic cells: lectin receptors, cell recognition and specific cell ligands. *In*: A. A. Moscona (ed.) *The Cell Surface in Development*, p. 67. (New York: John Wiley)

Nicolson, G. L. (1974). The interactions of lectins with animal cell surfaces. *Int. Rev. Cytol.*, **39**, 90

Okada, T. S., Eguchi, G. and Takeichi, M. (1973). The retention of differentiated properties by lens epithelial cells in clonal cell culture. *Dev. Biol.*, **34**, 321

Okada, T. S., Itoh, Y., Watanabe, K. and Eguchi, G. (1975). Differentiation of lens in cultures of neural retinal cells of chick embryos. *Dev. Biol.*, **45**, 318

O'Malley, B. W. and Means, A. R. (1974). Female steroid hormones and target cell nuclei. *Science*, **183**, 610

O'Malley, B. W., Woo, S. L. C., Harris, S. E., Rosen, J. M. and Means, A. R. (1975). Steroid hormone regulation of specific messenger RNA and protein synthesis in eukaryotic cells. *J. Cell. Physiol.*, **85**, 343

Paul, J. (1972). General theory of chromosome structure and gene activation in eukaryotes. *Nature (London)*, **238**, 444

Rabbitts, T. H. (1976). Bacterial cloning of plasmids carrying copies of rabbit globin messenger RNA. *Nature (London)*, **260**, 221

Rosbash, M., Campo, M. S. and Gummerson, K. S. (1975). Conservation of cytoplasmic poly (A)-containing RNA in mouse and rat. *Nature (London)*, **258**, 682

Stein, G. S., Spelsberg, T. C. and Kleinsmith, L. J. (1974). Nonhistone chromosomal proteins and gene regulation. *Science*, **183**, 817

Tiedemann, H. (1971). Extrinsic and intrinsic information transfer in early differentiation of amphibian embryos. *In: Control Mechanisms of Growth and Differentiation,* M. Balls and D. D. Davies (eds.) p. 233. (Cambridge: Cambridge University Press)

Tolstoshev, P., Mitchell, J., Lanyon, G., Williamson, R., Ottolenghi, S., Comi, P., Giglioni, B., Masera, G., Modell, B., Weatherall, D. J. and Clegg, J. B. (1976). Presence of gene for β globin in homozygous β_0 thalassaemia. *Nature (London)*, **259**, 95

Truman, D. E. S. (1974). *The Biochemistry of Cytodifferentiation.* (Oxford: Blackwell Scientific Publications)

Truman, D. E. S., Clayton, R. M., Gillies, A. G. and Mackenzie, H. J. (1976). RNA synthesis in the lenses of normal chicks and in two strains of chicks with hyperplasia of the lens epithelium. *Doc. Ophthalmol. Proceeding Series No.* **8**, 17

Wang, T. Y. and Nyberg, L. M. (1974). Androgen receptors in the nonhistone protein fractions of prostatic chromatin. *Int. Rev. Cytol.*, **39,** 1

Wessels, N. K. and Cohen, J. H. (1967). Early pancreas organogenesis: morphogenesis, tissue interactions and mass effects. *Dev. Biol.*, **15,** 237

Wolpert, L., Clarke, M. R. B. and Hornbruch, A. (1972). Positional signalling along *Hydra. Nature (London) New Biol.*, **239,** 101

Yaffe, D. (1969). Cellular aspects of muscle differentiation *in vitro. Curr. Top. Dev. Biol.*, **4,** 37

8

A Comparison of Leucocytes and Cultured 'Fibroblasts' in Diagnosis

R. A. HARKNESS

Department of Paediatric Biochemistry, Royal Hospital for Sick Children and University of Edinburgh, Edinburgh

A leucocyte is a differentiated specialized cell taken from the blood of a patient. The cultured cell described as a 'fibroblast' is a less differentiated cell grown, generally from skin or amniotic fluid cells, under precisely defined conditions in a laboratory. The uncultured leucocyte is therefore a more 'developed' cell and one which has been subject to many changes in its environment. Such changes can be shown to alter enzyme activities within the leucocyte.

The ready availability of peripheral blood, and the leucocytes which can be obtained from it, is a major factor determining the use of leucocytes in the early stages of the diagnosis of inherited metabolic disease. The cell generally used is the predominant leucocyte in human peripheral blood, the polymorphonuclear neutrophil leucocyte (PMNL). A simple fractionation method using dextran sedimentation is usually employed (Lehrer, 1971) yielding cells of which about 95% are PMNL, rather than the more elaborate separation of different cell types using columns of glass beads (Rabinowitz, 1973).

The next most abundant white cell in peripheral blood, the lymphocyte has been mostly studied by immunologists who have made extensive use of lymphocyte concentrates generally prepared by methods using Ficol (Boyum, 1968); again a variety of further separations often using surface properties are available. For example, B and T lymphocytes have been separated by centrifugation after rosette formation and used for studies of adenosine deaminase activity (Tung *et al.*, 1976), but such methods have been little used in diagnostic enzymology despite the fact that this cell is for a time in the newborn period the most abundant found in peripheral blood.

The ready availability of skin is a major factor determining its use as a starting point for the production of cultured cells. Such cells are similar to those obtained by culturing amniotic fluid cells, but culture is slow and expensive. The cell type produced is selected by its ability to grow *in vitro* and is often morphologically a 'fibroblast'. The advantage is the precise laboratory control of the environment of these cells. By standardizing the harvesting of cells after growth has become confluent, the laboratory procedure avoids problems connected with the changing enzymic composition of cells during growth. Cell culture methods are reviewed by Harnden (1977, p. 3 this volume) and the problem of control of cell density at confluence by Elsdale (1977, p. 33 this volume); cell density affects the yield and is therefore of some practical importance in diagnostic enzymology. In this chapter attention will therefore be concentrated mainly on leucocytes and the areas in which they contrast with cultured cells.

The need for studies of intracellular enzyme activities in the diagnosis of hereditary metabolic disease is relatively recent. Criteria which can be used in antenatal diagnosis of enzyme defects and in genetic counselling must be exact. Substrate accumulation is a useful method for deciding treatment as in amino acid or endocrine disorders but it may not be justified to assume from an abnormal substrate pattern, the precise nature of the enzyme defect. Such assumptions may apparently be justified from existing work and still be wrong (Harkness *et al.*, 1977). It is therefore necessary to check the diagnosis using enzyme estimations on cells from the affected family before attempting antenatal diagnosis.

For enzymological definition of a defect which is widely expressed, leucocytes are usually the first material studied. The highly specialized erythrocyte seems to contain only cytoplasmic enzymes but can be used for the study of cytoplasmic enzyme abnormalities where the relevant genetic locus controls that enzyme in affected tissues as well as in erythrocytes. This is the case with hypoxanthine–guanine phosphoribosyltransferase or HGPRT in the Lesch–Nyhan syndrome (Dancis *et al.*, 1973). For the investigation of defects in intracellular particles, in the lysosomes and related intracellular particles as well as in mitochondria, leucocytes have been extensively used. In general, defects in widely distributed systems such as the lysosomal acid hydrolases can be confirmed in 'fibroblasts' cultured from skin.

AN OVERALL SCHEME FOR DIAGNOSIS

The detailed and basic contributions to this symposium fit together into a scheme for the clinical investigation of hereditary metabolic disease (Figure 8.1). Appropriate parts of the scheme, shown in Figure 8.1 divided into clinical and laboratory sections, can be used to investigate individual clinical problems.

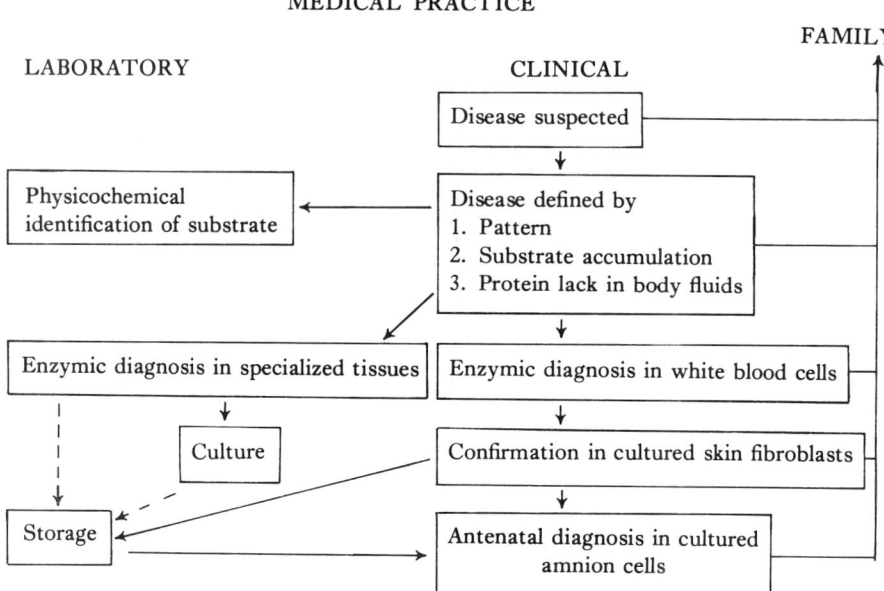

Figure 8.1   An overall scheme for the diagnosis of hereditary metabolic disease. The types of procedure are shown in boxes connected by continuous lines when they are part of accepted practice. Dotted lines indicate areas requiring research and development for useful clinical application. The column of procedures headed clinical, carry responsibility for the diagnosis and must be completed within time limits. Laboratory procedures are mainly concerned with scientific proof. The objective of such medical practice is to help an affected family

In the section headed 'clinical', laboratory workers and those in direct contact with the patient carry responsibility for any diagnoses which are made. Such work has also to be completed within time limits. The operations under the heading 'laboratory' do not suffer from such time limits and are often concerned with providing adequate scientific proof for conclusions. Since there are few effective treatments for defects in specialized tissues, enzymic diagnosis in specialized tissues is rarely an urgent problem. However, tissue stability is a major factor limiting the time for such enzymology and storage procedures are therefore important. In Figure 8.1 the solid lines joining the operations in the boxes are part of existing practice and the order in which the boxes are formed indicates the sequence of operations. The dotted lines joining the operations in the boxes indicate areas where research and development are needed to make such operations useful in solving medical problems. For example, culture of specialized tissues is still at an early stage (Dubowitz, 1977, p. 57 and Thompson, 1977, p. 69, this volume). Storage of some specialized tissues is still a problem; the polymorphonuclear neutrophil leucocyte, PMNL, commonly used in diagnosis is difficult to store (Pegg, 1976) whereas storage of lymphocytes and of cultured cells is standard practice (Pegg, 1977, p. 43, this volume).

The objective of the overall scheme of medical practice shown in Figure 8.1 is to help the affected family, a broader objective than helping one patient at one point in time.

LEUCOCYTES AND CULTURED 'FIBROBLASTS' AS UNDIFFERENTIATED CELLS

There appears to be a definite pattern in the evolution of diagnostic procedures for defects in generally distributed systems such as the lysosomes and this pattern is repeated in present practice (Figure 8.1). Historically, PMNL were first used for diagnosis, followed by the use of cells cultured from skin. There are exceptions, as in hypercholesterolaemia, where cultured cells were used extensively in early studies (Forbes–Robertson, 1977, p. 161, this volume), and PMNL are only now being used in diagnosis (Higgins, 1976). In general, defects diagnosable antenatally from cultured amnion cells can be diagnosed by enzyme assay in PMNL and confirmed in skin fibroblasts. Comparison of available lists of tissues suitable for the enzymic diagnosis of hereditary metabolic disease, for example that of Milunsky and Littlefield (1972), show very few examples in which PMNL and cultured skin fibroblasts cannot both be used; the discrepancies are not well documented. There are occasional difficulties in using PMNL which seem to stem from a few low activities in PMNL although these low activities are difficult to reproduce and are rarely recorded. These occasional low activities do not appear to be primarily due to methodological problems although sensitive methods are needed for both leucocytes and fibroblasts. As a consequence of the small amounts of material often available as well as low 'normal' activities of relevant enzymes, radioactively labelled substrates are often used (e.g. Brady, 1970).

Breakdown of cellular proteins: a factor controlling enzyme activities

The cause of the low activities occasionally found in PMNL may be, at least in part, the result of normal enzyme degradation in such a mature specialized end cell. The PMNL appears especially fitted for degrading bacterial components, and, apparently, itself. A variety of evidence indicates the importance of these degradation mechanisms in cells. In one family described by Dancis *et al.* (1973), very low levels of hypoxanthine–guanine phosphoribosyltransferase (HGPRT) activity were found in erythrocytes, probably due to their long life in the circulation when the short-lived leucocytes from the same patients had 10–15% of normal activity. However, there were abnormally rapid losses of HGPRT activity in these leucocytes despite storage at 4 °C. In another clinically affected family a marked reduction in erythrocyte HGPRT occurred probably due to storage at 4 °C, when activity in cultured skin fibroblasts was entirely normal (Adams and Harkness, unpublished observations). Even different enzymes with normal structures show a variety of stabilities; for example, phosphofructokinase is highly unstable in umbilical cord blood whereas enolase is stable (Kornazawa and Oski, 1975). The concentration of a substrate in a cell is a factor controlling the amount of a relevant enzyme persisting in a cell. For example, phosphoribosyl 1-pyrophosphate (PRPP) is believed to stabilize HGPRT in erythrocytes (Adams and Harkness, 1976). Tissue processing of proteins is itself under independent genetic control (Ganschow and Schimke, 1969) and varies with the tissue; tissue specific isoenzymes of adenosine deaminase are produced from a common structural unit (Hirschhorn, 1975).

Further evidence of tissue processing is provided by work on alkaline phosphatase levels in leucocytes; it can be shown that alkaline phosphatase activity is highest in the youngest cells (Williams, 1975). Such studies are possible because a histochemical method can give a good correlation with quantitative biochemical analyses (Diamant, Sadovsky and Gal, 1971). Further relevant evidence is available from studies on this enzyme; there is a single genetic locus controlling alkaline phosphatase in bone, liver and kidney as shown by the absence of the enzyme in these tissues in hypophosphatasia (Brydon et al., 1975). Therefore, the tissue specific isoenzymes of alkaline phosphatase from bone and liver are produced from a common structural unit by tissue specific processing.

The importance of variations in the environment affecting intracellular enzyme activities is shown by the large effects of hormones on leucocyte alkaline phosphatase (Rosner and Lee, 1965; Sadovsky et al., 1970). There may also be 'genetic' modification of hormonal response since the induction of alkaline phosphatase in leucocytes by glucocorticoids is more marked in Down's syndrome (McCoy and Ebadi, 1966). These and other findings suggest that a major factor determining measured enzyme activities may be the tissue processing of proteins within cells which might also be described as instability on storage.

Further support for the importance of catabolic breakdown of proteins as a factor governing measured activities is provided by early work on cells obtained from amniotic fluid. Enzyme activities in cells obtained directly from amniotic fluid showed some activities which were highly variable and difficult to detect (Sutcliffe and Brock, 1971, and others). Many enzyme activities are detectable in such cells (Nadler and Gerbie, 1969) but some low levels can be found when activities in corresponding cultured amniotic cells are normal (Butterworth et al., 1973). The problems of 'storage' in vivo or in vitro are avoided by the use of fresh samples of cultured cells.

Cultured cells: the persistent variability of enzyme activities

The use of cultured cells from skin and from amniotic fluid in the diagnosis of a variety of inherited metabolic diseases has been described by a number of authors and organizations (Nadler, 1969; Milunsky et al., 1970; Raine, 1972; Milunsky and Littlefield, 1972; Davidson and Rattazzi, 1972; Emery, 1973; Neufeld et al., 1975; Svennerholm, 1975). It is one of the functions of this volume to attempt to place such work in a broader context in which past achievements and some of the limitations of present practice may be seen.

Despite the relative constancy of in vitro conditions for cultured cells the problem of variability remains; this has been clearly defined in a series of studies on acid hydrolase activities (Galjaard et al., 1974). One of the factors affecting the metabolic behaviour of the cell is the influence of the culture medium which has already been reviewed (Rothblat and Cristofalo, 1972; Harnden, 1977, p. 3 and Lie, 1977, p. 16, this volume). Many other factors including genotype and hormones (Cox and MacLeod, 1964) have been recognized and attempts made to 'control' them, often with disappointing results. In some areas there remains a surprisingly good correlation between activities in cultured cells and disease processes.

Surprisingly there appears to be some inverse correlation between alkaline phosphatase activities in cultured cells and severity of the biological effects of hypophosphatasia in the limited data available (Brydon et al., 1975; Rattenbury et al., 1976). More evidence is available for HGPRT (Seegmiller, 1977, p. 187, this volume). However, in the mucopolysaccharidoses the correlation between the biological effects of the enzyme deficiency and its activity in cultured cells is poor (Spranger, 1975).

Variations in enzyme activity may be larger in the metabolically active cellular 'compartments', for example mitochondria and cytoplasm, than in the lysosomes, although little precisely comparable evidence exists. Wendel and his colleagues (1973) were able to diagnose maple syrup urine disease (MSUD) from assays of mitochondrial oxidative decarboxylation of branched chain α-keto acids but they were unable to distinguish heterozygotes possibly due to variations in the estimated activity despite the use of a sensitive radioisotopic assay method. They also found results from leucocytes more variable than from cultured skin fibroblasts. The enzymic diagnosis of propionicacidaemia involving mitochondrial propionyl-CoA carboxylase also appears to be difficult possibly due to variability in the estimated activities (Hsia et al., 1971; Gompertz, personal communication). In the area of mitochondrial function it therefore appears that less variability is encountered in cultured cells.

THE POLYMORPHONUCLEAR NEUTROPHIL LEUCOCYTE AS A DIFFERENTIATED CELL

The predominant leucocyte in peripheral blood except for a short time in the new born infant is the PMNL, a highly specialized cell with the function of ingesting and killing foreign microorganisms. PMNL are short lived cells; they have a half-life in the blood of 6–7 h and are therefore 'aging' rapidly. PMNL are sensitive to a large number of stimuli. After phagocytosis of foreign particles which results in a major redistribution of particulate enzyme activity (Stossel *et al.*, 1971) the PMNL responds with a burst of oxidative metabolism (Iyer *et al.*, 1961) and a related production of highly reactive oxidants like H_2O_2 which are used in bacterial killing (McRipley and Sbarra, 1967). In addition to the generally distributed lysosomes they contain a large number of specialized intracellular particles (Baggliolini *et al.*, 1970; Michell *et al.*, 1970).

The PMNL appears to be highly developed for the specialized work it does but in the course of killing other cells it appears to damage itself. For example, the burst of H_2O_2 production after phagocytosis will increase the breakdown of at least one enzyme metabolizing H_2O_2 since catalase is inactivated by H_2O_2 (Jones and Suggett, 1968). Such breakdown can be avoided by the addition of reducing agents (Holmes and Masters, 1959) or by the exclusion of oxygen (Cantz *et al.*, 1969). Thus, there is at least one biochemical mechanism explaining why the PMNL are difficult cells to handle in the laboratory in a way which gives reproducible high yields of enzyme activity.

Defects of the specialized function of PMNL are the subject of a large but separate literature some of which has been summarized (Bellanti and Dayton, 1975). Early work on microbicidal defects has been well reviewed (Klebanoff, 1971). The separate nature of these defects is emphasized by the lack of evidence for any impairment of host resistance in acid hydrolase deficiencies (Harkness, 1972) with the possible exception of the ill-understood Chediak–Higashi syndrome (Windhorst *et al.*, 1966). A more detailed treatment of the increasing number of inherited metabolic diseases involving the specialized function of PMNL in host resistance will be the subject of further reviews in this series of SSIEM monographs.

The object of this section is to show that the PMNL must be used for the diagnosis of hereditary metabolic disease of the specialized functions of the PMNL. The influence of some environmental factors will later be shown on the marker enzyme for myeloid differentiation, myeloperoxidase (MPO) itself involved in one microbicidal defect. This evidence suggests that in the MPO system environmental influences could easily be confused with genetic controls and result in the falsely positive diagnosis of hereditary metabolic disease.

The commonest recognized condition caused by a microbicidal defect in PMNL is chronic granulomatous disease of childhood (CGD). In this microbicidal defect, which in the commonest form is inherited as an X-linked recessive, it is generally agreed that there is a deficiency of nitroblue tetrazolium (NBT) reduction by the PMNL; defective NBT reduction is used to diagnose the condition. There is also a deficiency of H_2O_2 production after phagocytosis. The nature of the enzymic defect is controversial. Defects in a membrane NADH oxidase and in a particulate NADPH oxidase have been described (Segal, 1977). These may be reconcilable if both are involved in a cascade process and can function in parallel, possibly as a compensatory mechanism, as well as in series.

In one family with a rare autosomal recessively inherited form of CGD a glutathione peroxidase deficiency in PMNL has been described (Holmes et al., 1970). Another female variant form of CGD is known as Jobs syndrome (Bannatyne et al., 1969). Defective PMNL bactericidal activity has also been described in three sisters with lipochrome histiocytosis, a syndrome which includes rheumatoid arthritis (Rodey et al., 1969). A complete deficiency of leucocyte glucose-6-phosphate dehydrogenase is associated with a microbicidal defect whereas a partial deficiency with 25% of control activity was not associated with any defect in microbicidal activity (Cooper et al., 1970). A pyruvate kinase deficiency in leucocytes but not in erythrocytes has been found in a woman with an intracellular killing defect and repeated infections (Burge et al., 1976).

THE INFLUENCE OF ENVIRONMENTAL FACTORS ON ENZYME ACTIVITIES IN PMNL

The haem enzyme myeloperoxidase is closely associated with the microbicidal process and deficiencies of this enzyme are associated with a tendency towards an increased number of infections although these are surprisingly mild (Lehrer and Cline, 1969, and others). MPO is present in large amounts in PMNL (Evans and Rechcigl, 1967), is well characterized (Schultz and Shmukler, 1964) and is readily measurable (Klebanoff, 1965). It was therefore felt that MPO might be suitable for detecting and studying some environmental controls of an enzyme activity.

Reduction of the bactericidal activity of PMNL by oral ampicillin

A single oral dose of the usual therapeutic quantity of 500 mg of ampicillin significantly reduced PMNL bactericidal activity in healthy male volunteers (Raeburn, personal communication). There were significant enzymic changes consistent with this effect in that MPO activity fell and catalase activity rose. These results are summarized in Table 8.1.

Table 8.1   Effect of a single 500 mg dose of ampicillin on PMNL bactericidal, MPO and catalase activity

	Mean ± SEM Percentage of Control Values		
	(n = 4 or 5)		
Time in minutes after ampicillin	60	120	180
Activities in PMNL			
Bacterial Killing	28.3 ± 12.3	15.2 ± 4.3	80.4 ± 18.1
MPO	70 ± 7.8	81 ± 27	121 ± 35
Catalase	150 ± 22.1	153 ± 18.7	119 ± 48

Catalase and MPO values returned to normal in 24 h. For a detailed consideration of these findings and the relevant *in vitro* reduction of MPO activity by penicillin see Harkness and Grant (1977). Our results are consistent with earlier descriptions of leucocytes 'protecting' bacteria against the bactericidal actions of penicillin (Alexander and Good, 1968). It may also be relevant that a reduction of bacterial killing by *in vitro* incubation of leucocytes with a sulphonamide has been described (Lehrer, 1971), and is consistent with previously described inhibition of haem enzymes by sulphonamides (Main *et al.*, 1939; Lipmann, 1941).

The extent of these alterations in the PMNL with penicillin are surprising since penicillin is of low toxicity to cultured human cells (Byarugaba *et al.*, 1971). In addition, Mandell (1973) was unable to demonstrate significant amounts of penicillin inside PMNL. There is a great deal of evidence to show that these antibiotics do cross cell membranes; penicillins cross the gut wall and the placenta (MacCaulay *et al.*, 1966; Boreus, 1971). Penicillin has been detected in cultured cells (Eagle, 1954; Showacre *et al.*, 1961). In man, ampicillin has marked inhibitory effects on cytochrome P-450 dependent systems in liver (Harkness *et al.*, 1974) and placenta (Willman and Pulkinen, 1971). Since relatively small amounts of penicillin are retained in cells the haem enzymes would appear to be sensitive to these compounds and possibly to other 'amines' like sulphonamides. Studies of this problem are difficult due to the instability of penicillins in solution and to the slow reaction(s) of penicillins with haem enzymes (Renz *et al.*, 1972) and with other proteins (Corran and Waley, 1975).

In view of the widespread use of penicillin in tissue preparation for culture and in culture media (Kruse and Patterson, 1973) further studies designed to detect penicillin effects on cultured cells, especially on haem enzyme dependent systems, may be needed.

One immediate and direct consequence of these results is to emphasize that environmental factors can alter enzyme activities in PMNL. It will now be necessary carefully to exclude penicillin effects before accepting the primary effect of trauma and acute infections on PMNL bactericidal activity (Drutz and Cline, 1976) and the effect of virus infections in children (Craft *et al.*, 1976).

Nutritional effects on PMNL function are well documented, for example iron deficiency with anaemia (Chandra, 1973) or in a periodic fashion, without anaemia (Harkness and Grant, 1977). Vitamin deficiencies are also important; cobalamin deficiency impairs propionate oxidation and serine biosynthesis (Robertson *et al.*, 1976). Generalized undernutrition causes a spectrum of changes (Avila *et al.*, 1973) and a reversible killing defect has been described in anorexia nervosa (Gotch *et al.*, 1975). Such environmental factors are usually variable and this may help to distinguish between defective genetic control and an environmental effect in enzymic diagnosis.

This section of the review is entitled the polymorphonuclear neutrophil leucocyte as a differentiated cell. The morphological and functional differences make it clear that the lymphocyte is a different cell from the PMNL. The lymphocyte also reacts differently to drugs (McCurrach *et al.*, 1970), although the difference is sometimes forgotten. An overall review of the biochemical differences between the two cell types is needed but the comparisons may be difficult. Even adenosine deaminase activities which appear to be related to lymphocyte function are not higher in lymphocytes unless a comparison is made between lymphocytes and PMNL from the same individual thus eliminating genetic differences. Immunologists have contributed so much to our knowledge of lymphocyte function that it is surprising that more is not known about the biochemistry of these cells.

THE FIBROBLAST AS A DIFFERENTIATED CELL

The main cell type grown from skin or from amnion cells is fibroblast-like and relatively undifferentiated. In many enzymic diagnoses it is used as an undifferentiated cell having basic cellular mechanisms. However, such cells can show the specialized functions of fibroblasts, for example collagen production (Grant, 1975; Bailey, 1975).

There is also evidence that skin samples from different sites can behave differently. Only genital skin on culture gives rise to 'fibroblasts' in which a reduction of steroid 5α-reductase activity is found in a form of defective male development, familial incomplete male pseudohermaphroditism, type 2 (Moore *et al.*, 1975). Steroid 5α-reductase activity was very low in extracts from cultured cells and was highly variable; control values from genital skin cultures varied almost a hundredfold. This variation was not due to low substrate concentration or to variations in the number of cell passages which only account for an approximately twofold increase between the 10th and 20th transfers. This is further evidence of the problem of variation in activities in cultured cells.

ISOENZYMES WITH DIFFERENT GENETIC CONTROLS

Both the PMNL and the cultured fibroblast are used as models when the clinically significant changes occur in other tissues. In such applications, there is often an assumption that the enzyme activity measured in the model cell is the same as that in the affected tissue. This assumption may not be justified. For example, the enzyme converting androstenedione to testosterone in testis, 17β-hydroxysteroid oxidoreductase has a different genetic control from the generally distributed 17β-hydroxysteroid oxidoreductase most studied in erythrocytes (Harkness *et al.*, 1975). It is therefore not possible to use cells from blood to diagnose the enzyme defect in gonadal 17β-hydroxysteroid oxidoreductase deficiency; a condition which causes an almost complete failure of masculinization in genetic males. Although such a difference might have been predicted, it may be unwise to generalize. There is a surprising degree of correlation between alkaline phosphatase activities in cultured cells and severity of disease in the limited data available (Brydon et al., 1975; Rattenbury et al., 1976), despite the existence of tissue-specific processing and independent genetic control by at least three loci. For diagnostic purposes it appears advisable to show that the enzyme activity measured in the easily available PMNL and cultured fibroblasts is under the same genetic control as the enzyme in the affected tissue. Ideally these investigations should form part of the early descriptions of a hereditary metabolic disease.

The relevance of such primary isoenzymes to hereditary metabolic disease or to physiological or other pathological processes is unclear apart from their nuisance value in diagnostic work. The 21-hydroxylation of steroids by cultured fibroblasts most readily shown in the S-phase of the cell cycle (Berliner and Garzon, 1969) has not predictably been used in diagnostic work on the common hereditary metabolic disease due to defective 21-hydroxylation of steroids. These observations may be relevant because after more than 20 years work on such patients it is still not clear why the defect is always 'partial'. Some steroid 21-hydroxylation persists and with negative feedback controls ensures normal levels of the end product of the affected biosynthetic chain, cortisol. It seems justifiable to suggest that such a low but widely distributed capacity for the 21-hydroxylation of steroids could function in the presence of the higher substrate concentrations which occur when the specialized system in the adrenal cortex is ineffective. Similar parallel compensatory pathways appear to operate in the bactericidal capacity of PMNL in which similar increases of substrate, in this case H_2O_2, can also be shown (Klebanoff and Pincus, 1971).

Redundancy in biological systems is common. Such redundancy offers an important safety mechanism although making diagnostic work, and an understanding of biological systems, more difficult.

References

Adams, A. and Harkness, R. A. (1976). Developmental changes in purine phosphoribosyltransferases. *Biochem. J.*, **160**, 565

Alexander, J. W. and Good, R. A. (1968). Effect of antibiotics on the bactericidal activity of human leucocytes. *J. Lab. Clin. Med.*, **71**, 971

Avila, J. L., Velazquez, G., Correa, A. C., Correa, C., Castillo, C. and Convit, J. (1973). Leukocytic enzyme differences between the clinical forms of malnutrition. *Clin. Chim. Acta*, **49**, 5

Baggliolini, M., Hirsch, J. G. and de Duve, C. (1970). Further biochemical and morphological studies of granule fractions from rabbit heterophil leukocytes. *J. Cell Biol.*, **45**, 586

Bailey, A. J. (1975). Biosynthesis of collagen cross-links: Relationship of hereditable disorders. *In:* J. B. Holton and J. T. Ireland (eds.), *Inborn Errors of Skin, Hair and Connective Tissue*, p. 105. (Lancaster: MTP Press Limited)

Bannatyne, R. M., Skowron, P. N. and Weber, J. L. (1969). Job's syndrome: a variant of chronic granulomatous disease. *J. Pediatr.*, **75**, 236

Bellanti, J. A. and Dayton, D. H. (1975). *The Phagocytic Cell in Host Resistance.* (New York: Raven Press)

Berliner, D. L. and Garzon, P. (1969). Steroid 21-hydroxylase activity by fibroblasts during the cell cycle. *Steroids*, **14**, 409

Boreus, L. O. (1971). Placental transfer of ampicillin in man. *Acta Pharmacol. Toxicol.*, **29**, Suppl. 3, 250

Boyum, A. (1968). Separation of leucocytes from blood and bone marrow. *Scand. J. Clin. Lab. Invest.*, **21**, Suppl. 97

Brady, R. O. (1970). Prenatal diagnosis of lipid storage diseases. *Clin. Chem.*, **16**, 811

Brydon, W. G., Crofton, P. M., Smith, A. F., Barr, D. G. D. and Harkness, R. A. (1975). Hypophosphatasia: enzyme studies in cultured cells and tissues. *Biochem. Soc. Trans.*, **3**, 927

Burge, P. S., Johnson, W. A. and Hayward, A. R. (1976). Neutrophil pyruvate kinase deficiency with recurrent staphylococcal infections, first reported case. *Brit. Med. J.*, **1**, 742

Butterworth, J., Sutherland, J. R., Broadhead, D. M. and Bain, A. D. (1973). Lysosomal enzymes in cultured amniotic fluid cells. *Clin. Chim. Acta*, **44**, 295

Byarugaba, W., Rudiger, H. W., Koske-Westphal, T., Wohler, W. and Passarge, E. (1975). Toxicity of antibiotics on cultured human skin fibroblasts. *Humangenetik*, **28**, 263

Cantz, M., Morikofer-Zweis, S., Bossi, E., Kaufmann, H., von Wartburg, J. P. and Aebi, H. (1968). Alternative molecular forms of erythrocyte catalase. *Experientia*, **24**, 119

Chandra, R. K. (1973). Reduced bactericidal capacity of polymorphs in iron deficiency. *Arch. Dis. Child.*, **48**, 864

Cooper, M. R., DeChatelet, L. R., McCall, C. E., Lavia, M. F., Spurr, C. L. and Baehner, R. L. (1970). Leucocyte G-6-PD deficiency. *Lancet*, **ii**, 110

Corran, P. H. and Waley, S. G. (1975). The reaction of penicillin with proteins. *Biochem. J.*, **149**, 357

Cox, R. P. and MacLeod, C. M. (1964). Regulation of alkaline phosphatase in human cell cultures. *Cold Spring Harbor Symp. Quant. Biol.*, **29**, 233

Craft, A. W., Reid, M. M. and Low, W. T. (1976). Effect of virus infections on polymorph function in children. *Br. Med. J.*, **1**, 1570

Dancis, T., Yip, L. C., Cox, R. P., Piomelli, S. and Balis, M. E. (1973). Disparate enzyme activity in erythrocytes and leukocytes: a variant of hypoxanthine–guanine phosphoribosyltransferase deficiency with an unstable enzyme. *J. Clin. Invest.*, **52**, 2068

Davidson, R. G. and Rattazzi, M. C. (1972). Prenatal diagnosis of genetic disorders; trials and tribulations. *Clin. Chem.*, **18**, 179

Diamant, Y. S., Sadovsky, E. and Gal, A. (1971). Evaluation of the leukocyte alkaline phosphatase activity. Comparison of the biochemical and cytochemical assay. *Clin. Chim. Acta*, **34**, 73

Drutz, D. J. and Cline, M. J. (1975). Intermittent neutrophil-monocyte bactericidal defects in a patient with sarcoidosis. *Am. Rev. Resp. Dis.*, **112**, 387

Eagle, H. (1954). The binding of penicillin in relation to its cytotoxic action. III. The binding of penicillin by mammalian cells in tissue culture. *J. Exp. Med.*, **100**, 117

Emery, A. E. G. (1973). *Antenatal Diagnosis of Genetic Disease*. (London and Edinburgh: Churchill-Livingstone)

Evans, W. H. and Rechcigl, M. (1967). Factors influencing myeloperoxidase and catalase activities in polymorphonuclear leukocytes. *Biochim. Biophys. Acta*, **148**, 243

Galjaard, Reuser, A. J. J., Heukels-Dully, M. J., Hoogeveen, A., Keijzer, W., de Wit-Verbeek, H. A. and Niermeijer, M. F. (1974). Genetic heterogeneity and variation of lysosomal enzyme activities in cultured human cells. *In:* J. M. Tager, G. J. M. Hoogwinkel and W. Th. Daems (eds.), *Enzyme Therapy in Lysosomal Storage Disease*, p. 35. (Amsterdam: North-Holland Publishing Co.)

Ganschow, R. E. and Schimke, R. T. (1960). Independent genetic control of the catalytic activity and rate of degradation of catalase in mice. *J. Biol. Chem.*, **244**, 4649

Gotch, F. M., Spry, C. J. F., Mowat, A. G., Beeson, P. B. and MacLennan, I. C. M. (1975). Reversible granulocyte killing defect in anorexia nervosa. *Clin. Exp. Immunol.*, **21**, 244

Grant, M. E. (1975). Structure and biosynthesis of collagen. *In:* J. B. Holton and J. T. Ireland (eds.), *Inborn Errors of Skin, Hair and Connective Tissue,* p. 83. (Lancaster: MTP Press Limited)

Harkness, R. A. (1972). A search for some enzymic basis for host resistance: the effect of isolated enzyme defects. *In:* T. MacPhee (ed.), *Host Resistance to Commensal Bacteria,* p. 209. (London and Edinburgh: Churchill-Livingstone)

Harkness, R. A., Scott, R. D. M. and Strong, J. A. (1974). Physiological and pharmacological factors affecting 6β-hydroxylation and 17-epimerisation of methandrostenolone. *Biochem. Soc. Trans.,* **2,** 119

Harkness, R. A., Thistlethwaite, D., Darling, J. A. B., Skakkebaek, N. E. and Corker, C. S. (1975). 17β-hydroxysteroid oxidoreductase deficiency causing male pseudohermaphroditism. *J. Endocrinol.,* **67,** 16P

Harkness, R. A. and Grant, M. (1977). Pharmacological and other controls of enzyme activities in polymorphonuclear phagocytes. *In:* *Proceedings of the First European Conference on Phagocytic Leucocytes.* (In press)

Harkness, R. A., Cockburn, F., Grant, M., Giles, M. M., Turner, T. L. and Darling, J. A. B. (1977). A new variety of maple syrup urine disease. *Ann. Clin. Biochem.,* **14,** 146

Higgins, M. J. P. (1976). The regulation of cholesterol metabolism. *Biochem. Soc. Trans.,* **4,** 572

Hirschhorn, R. (1975). Conversion of human erythrocyte adenosine deaminase activity to different tissue specific isoenzymes. Evidence for a common catalytic unit. *J. Clin. Invest.,* **55,** 661

Holmes, R. S. and Masters, C. J. (1959). On the tissue and subcellular distribution of multiple forms of catalase in the rat. *Biochem. Biophys. Acta,* **191,** 488

Holmes, B., Park, B. H., Malawista, S. E., Quie, P. G., Nelson, D. L. and Good, R. A. (1970). Chronic granulomatous disease in females. A deficiency of leucocyte glutathione peroxidase. *N. Engl. J. Med.,* **283,** 217

Hsia, Y. E., Scully, K. J. and Rosenberg, L. E. (1971). Inherited propionyl-CoA carboxylase deficiency in 'ketotic hyperlgycinemia'. *J. Clin. Invest.,* **50,** 127

Iyer, G. Y. N., Islam, D. M. F. and Quastel, J. H. (1961). Biochemical aspects of phagocytosis. *Nature (London),* **192,** 535

Jones, P. and Suggett, A. (1968). The catalase–hydrogen peroxide system. Kinetics of catalatic action at high substrate concentrations. *Biochem. J.,* **110,** 617

Klebanoff, S. J. (1965). Inactivation of estrogen by rat uterine preparations. *Endocrinology,* **76,** 301

Klebanoff, S. J. (1971). Intraleukocytic microbicidal defects. *Annu. Rev. Med.,* **22,** 39

Klebanoff, S. J. and Pincus, S. H. (1971). Hydrogen peroxide utilization in myeloperoxidase deficient leukocytes: a possible microbicidal control mechanism. *J. Clin. Invest.,* **50,** 2226

Kornazawa, M. and Oski, F. A. (1975). Biochemical characteristics of young and old erythrocytes of the newborn infant. *J. Pediatr.,* **87,** 102

Kruse, P. F. and Patterson, M. K. (1973). *Tissue Culture: Methods and Applications.* (New York and London: Academic Press)

Lehrer, R. I. and Cline, M. J. (1969). Leukocyte myeloperoxidase deficiency and disseminated candidiasis: the role of myeloperoxidase in resistance to candida infection. *J. Clin. Invest.,* **48,** 1478

Lehrer, R. I. (1971). Inhibition by sulfonamides of the candidacidal activity of human neutrophils. *J. Clin. Invest.,* **50,** 2498

Lipmann, F. (1941). The oxidation of *p*-aminobenzoic acid catalyzed by peroxidase, and its inhibition by sulphanilamide. *J. Biol. Chem.,* **139,** 977

MacAuley, M. A., Abou-Sabe, M. and Charles, D. (1966). Placental transfer of ampicillin. *Am. J. Obstet. Gynecol.,* **96,** 943

Main, E. R., Shinn, L. E. and Mellon, R. R. (1939). Anticatalase activity of sulphanilamide and related compounds. IV. Peroxide accumulation and growth inhibition of pneumococcus. *Proc.*

Soc. Exp. Biol. Med., **42**, 115

Mandell, G. L. (1973). Interaction of intraleukocytic bacteria and antibiotics. *J. Clin. Invest.*, **52**, 1673

McCoy, E. and Ebadi, M. (1966). Induction of leukocyte alkaline phosphatase in Down's syndrome. *J. Pediatr.*, **68**, 835

McCurrach, P. H., Park, J. K. and Perry, W. L. M. (1970). The effect of drugs which induce agranulocytosis on the metabolism of separated human polymorphonuclear leucocytes and lymphocytes. *Br. J. Pharmacol.*, **38**, 608

McRipley, R. J. and Sbarra, A. J. (1967). Role of the phagocyte in host–parasite relationships. XI. Relationship between stimulated oxidative metabolism and hydrogen peroxide formation and intracellular killing. *J. Bacteriol.*, **94**, 1417

Michell, R. H., Karnovsky, M. J. and Karnovsky, M. L. (1970). The distribution of some granule associated enzymes in guinea pig polymorphonuclear leucocytes. *Biochem. J.*, **116**, 207

Moore, R. J., Griffin, J. E. and Wilson, J. D. (1975). Diminished 5α-reductase activity in extracts of fibroblasts cultured from patients with familial incomplete male pseudohemaphroditism Type 2. *J. Biol. Chem.*, **250**, 7168

Milunsky, A., Littlefield, J. W., Kanfer, J. N., Kolodny, E. H., Shih, V. E. and Atkins, L. (1970). Prenatal genetic diagnosis. *N. Engl. J. Med.*, **283**, 1370, 1441 and 1498

Milunsky, A. and Littlefield, J. W. (1972). The prenatal diagnosis of inborn errors of metabolism. *Annu. Rev. Med.*, **23**, 57

Nadler, H. L. (1969). Prenatal detection of genetic defects. *J. Pediatr.*, **74**, 132

Nadler, H. L. and Gerbie, A. B. (1969). Enzymes in noncultured amniotic fluid cells. *Am. J. Obstet. Gynecol.*, **103**, 710

Neufeld, E. F., Lim, J. W. and Sharpiro, L. J. (1975). Inherited disorders of lysosomal metabolism. *Annu. Rev. Biochem.*, **44**, 357

Pegg, D. E. (1976). Long-term preservation of cells and tissues: a review. *J. Clin. Pathol.*, **29**, 271

Rabinowitz, Y. (1973). Nonenzymatic dissociations. A. Leukocyte cell separations on glass. *In*: P. F. Kruse and M. K. Patterson (eds.), *Tissue Culture: Methods and Applications*, p. 25. (New York and London: Academic Press)

Raine, D. N. (1972). Management of inherited metabolic disease. *Br. Med. J.*, **1**, 329

Rattenbury, J. M., Blau, K., Sandler, M., Pryse-Davies, J., Clark, P. J. and Pooley, S. S. F. (1976). Prenatal diagnosis of hypophosphatasia. *Lancet*, **i**, 306

Renz, M., Nicol, A. D. and Harkness, R. A. (1972). Drugs as inhibitors of catalase and peroxidase. *In*: T. MacPhee (ed.), *Host Resistance to Commensal Bacteria*, p. 209. (London and Edinburgh: Churchill-Livingstone)

Robertson, J. S., Hsia, Y. E. and Scully, K. J. (1976). Defective leukocyte metabolism in human cobalamin deficiency: impaired propionate oxidation and serine biosynthesis reversible by cyanocobalamin therapy. *J. Lab. Clin. Med.*, **87**, 89

Rodey, G. E., Park, B. H., Ford, D. K., Gray, B. H. and Good, R. A. (1970). Defective bactericidal activity of peripheral blood leukocytes in lipochrome histiocytosis. *Am. J. Med.*, **49**, 322

Rosner, F. and Lee, S. L. (1965). Endocrine relationships of leukocyte alkaline phosphatase. *Blood*, **25**, 356

Rothblat, G. H. and Cristofalo, V. J. (1972). *Growth, Nutrition and Metabolism of Cells in Culture*. Vols. I and II (New York and London: Academic Press)

Sadovsky, E., Zuckerman, H., Diamant, Y. Z. and Polishuk, W. Z. (1970). Leukocyte alkaline phosphatase and fetal prognosis in placental dysfunction. *Am. J. Obstet. Gynecol.*, **108**, 979

Segal, A. (1977). Localisation of enzymes in neutrophils involved in the microbicidal process. The enzyme defect in chromic granulomatous disease. *In*: *Proceedings of the First European Conference on Phagocytic Leukocytes*. (In press)

Schultz, J. and Shmukler, H. W. (1964). Myeloperoxidase of the leukocyte of normal human blood. II. Isolation, spectrophotometry and amino acid analysis. *Biochemistry,* **3,** 1234

Showacre, J. L., Hopps, H. E., du Buy, H. G. and Smadel, J. E. (1961). Effects of antibiotics on intracellular *Salmonella typhosa.* I. Demonstration by phase microscopy of prompt inhibition of intracellular multiplication. *J. Immunol.,* **87,** 153

Spranger, J. (1975). Morphological aspects of the mucopolysaccharidoses. *In:* J. B. Holton and J. T. Ireland (eds), *Inborn Errors of Skin, Hair and Connective Tissue,* p. 39. (Lancaster: MTP Press Limited)

Stossel, T. P., Pollard, T. D., Mason, R. J. and Vaughan, M. (1971). Isolation and properties of phagocytic vesicles from polymorphonuclear leucocytes. *J. Clin. Invest.,* **50,** 1745

Sutcliffe, R. G. and Brock, D. J. H. (1971). Enzymes in uncultured amniotic fluid cells. *Clin. Chim. Acata,* **31,** 363

Svennerholm, L. (1975). List of European laboratories which perform prenatal diagnosis of inherited metabolic disorders. S-422 03 Hisings Backa 3, Sweden

Tung, R., Silber, R., Quagliata, F., Conklym, M., Gottesman, J. and Hirschhorn, R. (1976). Adenosine deaminase activity in chronic lymphocytic leukaemia. Relationship to B- and T-cell subpopulations. *J. Clin. Invest.,* **57,** 756

Wendel, C., Wohler, W., Goedde, H. W., Langenbeck, U., Passarge, E. and Rudiger, H. W. (1973). Rapid diagnosis of maple syrup urine disease (branched chain ketoaciduria) by microenzyme assay in leukocytes and fibroblasts. *Clin. Chim. Acta,* **45,** 433

Williams, D. M. (1975). Leucocyte alkaline phosphatase as a marker of cell maturity: a quantitative cytochemical and autoradiographic study. *Br. J. Haematol.,* **31,** 371

Willman, R. and Pulkinen, M. O. (1971). Reduced maternal plasma and urinary oestriol during ampicillin treatment. *Am. J. Obstr. Gynecol.,* **109,** 873

Windhorst, D. B., Zelickson, A. S. and Good, R. A. (1966). Chediak–Higashi syndrome: hereditary gigantism of cytoplasmic organelles. *Science,* **151,** 81

9
Acid Hydrolase Production, Release and Uptake by Cultured Fibroblasts

K. VON FIGURA

Institute of Physiological Chemistry, University of Münster, West Germany

Cultivated skin fibroblasts have become a major tool for the biochemical investigation of lysosomal storage diseases, since fibroblasts grown from the skin of patients affected with one of these genetic disorders express the basic defect in cell culture (Hers and van Hoof, 1973). The elucidation of the various enzyme defects during the last decade has facilitated the genotype specific classification and diagnosis of these diseases, and has provided the basis for prenatal diagnosis and genetic counselling. Moreover, promising *in vitro* models for specific enzyme replacement therapy have been developed using cultivated skin fibroblasts as a test system. These investigations have extended our basic knowledge of the physiological function of lysosomal hydrolases in the degradation of macromolecules such as mucopolysaccharides and sphingolipids.

FORMATION OF LYSOSOMAL HYDROLASES

In contrast to the progress made in the characterization of lysosomal enzymes of cultivated fibroblasts, little is known about their biosynthesis, the packaging of lysosomal hydrolases into vesicles and the subsequent fusion of these vesicles with either substrate-containing vacuoles or the cell membrane. The biochemical approach to these processes has been hampered by the difficulty in obtaining well-defined portions of the various subcellular structures in quantities that would allow detailed biochemical analyses.

The greater part, if not all lysosomal hydrolases are glycoproteins (Goldstone and Koenig, 1970). It is generally accepted that the protein moieties of glycoproteins are synthesized on the ribosomes of the rough endoplasmic reticulum, then transferred into the cisternal space of the endoplasmic reticulum and subsequently transported to the Golgi apparatus (de Duve and Wattiaux, 1966; Cohn and Fedorko, 1969). This mechanism is similar to that described for the secretory proteins in pancreatic acinar cells (Caro and Palade, 1969; Jamieson and Palade, 1967a; Jamieson and Palade, 1967b). Glycosylation is thought to take place during the transport of the proteins along the membranes of the rough and smooth endoplasmic reticulum and in the Golgi apparatus. This concept is based mainly on ultrastructural cytochemical and autoradiographic observations made on a variety of cell types, such as hepatocytes, neurons, collecting tubule cells of the kidney, polymorphonuclear leucocytes and peritoneal macrophages, using acid phosphatase as a histochemical marker enzyme for lysosomal hydrolases (Cohn and Fedorko, 1969; Ericsson, 1969; Koenig, 1969; Maunsbach, 1969; Holtzmann, 1969; Novikoff, 1967).

Biochemical evidence for the synthesis of lysosomal enzymes on the rough endoplasmic reticulum and subsequent transfer to the Golgi apparatus via the smooth endoplasmic reticulum has been presented for β-glucuronidase, acid phosphatase, β-galactosidase, β-N-acteylhexosaminidase and arylsulfatase A (Walkinshaw and van Lanckner, 1964; van Lanckner and Lentz, 1970; Ide and Fishman, 1969; Kato et al., 1970; Kato et al., 1972; Novikoff et al., 1964; Nichols et al., 1971; Goldstone and Koenig, 1973a). Studies on subcellular fractions from rat kidney indicate that lysosomal hydrolases are synthesized on restricted portions of the rough endoplasmic reticulum, where they occur as bound forms with rather basic isoelectric points (pI > 6.0). As the nascent basic forms migrate to the smooth endoplasmic reticulum and the Golgi apparatus they become more soluble and acidic by attachment of terminal neuraminic acid residues to the carbohydrate moieties (Goldstone and Koenig, 1973b).

Some controversy exists about the precise site of packaging of lysosomal hydrolases into the membrane-surrounded vesicles called primary lysosomes, i.e. lysosomes whose enzymes have not yet participated in degradative processes of any kind. Whereas in earlier studies the peripheral portions of the Golgi apparatus have been considered as the site where vesicles containing acid hydrolyases are being budded off (Cohn and Fedorko, 1969; Ericsson, 1969; Koenig, 1969), Novikoff and associates (Novikoff et al., 1964; Novikoff, 1967) have proposed the GERL concept (abbreviated from Golgi, endoplasmic reticulum and lysosomes). According to the GERL concept vesicles containing acid hydrolases arise from a special portion of the smooth endoplasmic reticulum which is in close association with the concave face of the Golgi apparatus (emitting face) and which consists of small cisternal portions and anastomosing tubules.

As pointed out earlier, the studies which led to the two concepts described for the formation of primary lysosomes were made on cell types other than fibroblasts. Studies on cultivated skin fibroblasts may well modify this concept, since the formation of secondary lysosomes in cultivated fibroblasts seems to differ from that in other cell types.

SECRETION OF LYSOSOMAL ENZYMES

Lysosomal enzymes found in the culture medium of fibroblasts can derive either from viable cells or from non-viable or severely damaged cells. The term secretion is restricted to a release of lysosomal enzymes which is not accompanied by a release of enzymes derived from other cell compartments, such as cytoplasm. Thus, secretion is a characteristic of viable cells, whereas the release of enzymes from other cell compartments is a sign of grossly altered cell permeability or cell death (Dingle, 1969).

Secretion of lysosomal enzymes requires intact cellular metabolism for the cell to form vesicles filled with acid hydrolases, to transport these vesicles from the site of formation to the cell membrane and to induce fusion of the vesicles with the cell membrane.

The medium surrounding cultivated skin fibroblasts contains lysosomal enzymes secreted by the cells. This has been verified for a number of lysosomal enzymes. With the exception of β-hexosaminidase the enzyme activities released by the cells are very low and can hardly be detected without concentration of the medium. β-Hexosaminidase has therefore been used as marker enzyme for secretion of lysosomal hydrolases. In the first four hours after feeding with fresh medium, containing 10% heat inactivated fetal calf serum, fibroblasts release about 2–3% of their total β-hexosaminidase activity into the medium. After this initial phase of rapid release the cells secrete about 1–3% of the total β-hexosaminidase per 24 h at a fairly constant rate into the medium (Figure 9.1). During the first four hours the cells are completely resistant to new stimulation of the secretion by refeeding. After 24 h, however, a rapid release can be again provoked by addition of fresh medium. Feeding with conditioned medium leads to a much lower release of lysosomal enzymes. The time course of β-hexosaminidase release is identical if fibroblasts are fed with a chemically defined medium without serum. Therefore it seems likely that feeding with unconditioned medium provides a stimulus to the cells for increased lysosomal enzyme secretion and that cells become refractory to this stimulus for at least four hours.

The rate of secretion seems to be controlled by the integrity of microfilaments and microtubules. Cytochalasin B, a mould metabolite, produces a disintegration of the microfilaments in a number of different cell types (Wessels et al., 1971). Microfilaments seem to consist of actin (Lengsfeld et al., 1974). Cultivation of skin fibroblasts in the presence of cytochalasin B is followed by an increased secretion of β-hexosaminidase without concomitant release of lactate dehydrogenease (Figure 9.2). Not all the effects of cytochalasin B can be related to its effect on the microfilaments (Kletzien et al.,

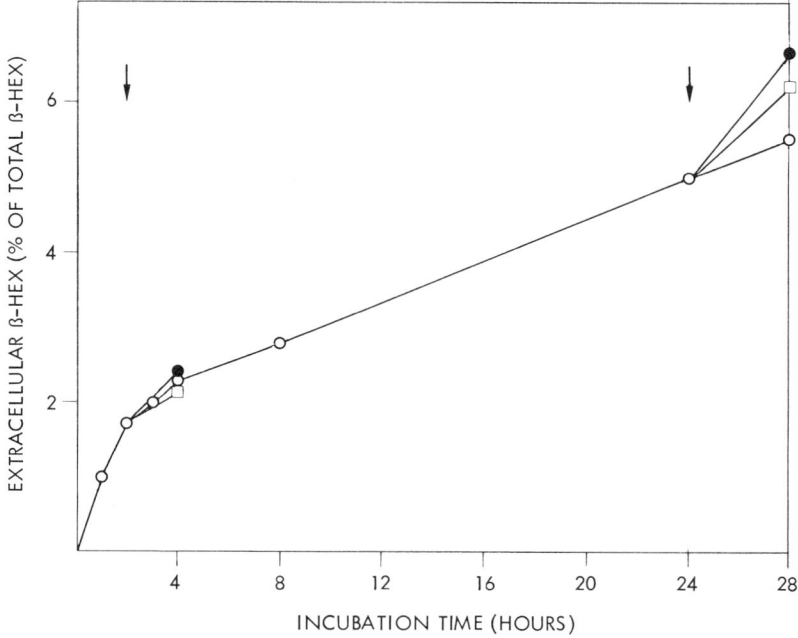

↓ REFEEDING WITH FRESH MEDIUM (●)   REFEEDING WITH CONDITIONED MEDIUM (□)

Figure 9.1   Time course of β-hexosaminidase secretion by cultivated skin fibroblasts. After 2 and 24 h parallel cultures were fed with either fresh medium (●) or conditioned medium (□). Secretion is expressed as extracellular β-hexosaminidase activity as a percentage of total intra- and extra-cellular β-hexosaminidase activity

1972; Estensen and Plagemann, 1972). As long as the precise mode of action of cytochalasin B in unknown, the conclusion that microfilaments are involved in the control of secretion must be viewed with caution. Antimicrotubular drugs, such as colchicine, vinblastine and vincristine, which cause a depolymerization of the microtubules, stimulate the secretion of β-hexosaminidase more than six-fold over a wide range of concentrations (Figure 9.3). This stimulatory effect of antimicrotubular agents on the secretion is specific for lysosomal hydrolases. In fibroblasts the secretion of collagen (Dehm and Prockop, 1972) is inhibited and that of proteoglycans is not influenced by antimicrotubular drugs (von Figura, unpublished).

Preincubation of fibroblasts with the antimalarial drug chloroquine leads to release of lysosomal hydrolases into the medium and a decrease of intracellular lysosomal enzyme activities (Wiesmann *et al.*, 1975). These data were interpreted as showing competition of the drug for lysosomal enzyme binding sites on the membranes of the lysosomes.

In comparison to polymorphonuclear leucocytes where a number of factors, including vitamin A, antigen–antibody complexes, complement C3 and C5a,

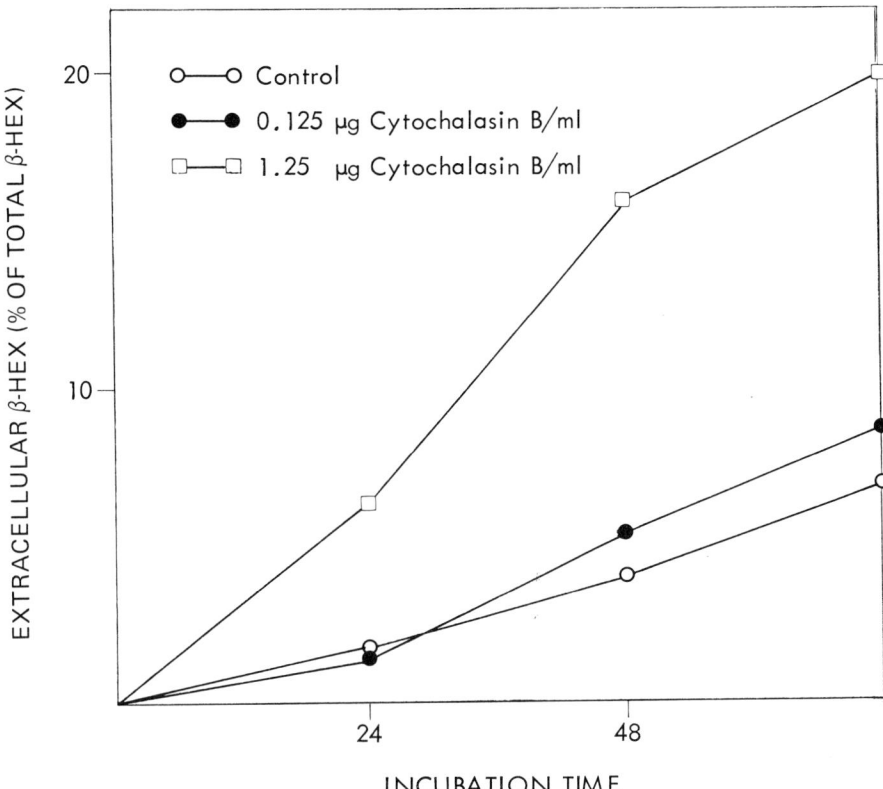

Figure 9.2 Effect of cytochalasin B on the secretion of β-hexosaminidase. Twenty-four hours after feeding the drug was added and the enzyme release was measured after various periods of incubation. Controls contained the solvent of cytochalasin B, dimethylsulphoxide ($2.5\,\mu l$/ml medium)

cyclic AMP, cyclic GMP, prostaglandins, cholera endotoxin, histamine and urate crystals, are known to influence the secretion of lysosomal enzymes, our knowledge about factors controlling the secretion of lysosomal enzymes in fibroblasts is rather limited. Whereas fairly well established concepts have been presented for the role of secretion and extracellular function of lysosomal enzymes by polymorphonuclear leucocytes and tissues such as cartilage and bone, the biological significance of secretion of lysosomal hydrolases by cultured 'fibroblasts' remains to be elaborated.

UPTAKE OF LYSOSOMAL HYDROLASES

Lysosomal hydrolases added to a culture medium of skin fibroblasts can be taken up by the cells. The uptake of a particular acid hydrolase can easily be followed if fibroblasts deficient in that enzyme are used. Moreover, such enzyme deficient fibroblasts allow the study of the metabolic effect of the ingested enzyme. The correction of the impaired catabolism of that material, which had been stored due to the enzyme deficiency, indicates that after it has been taken

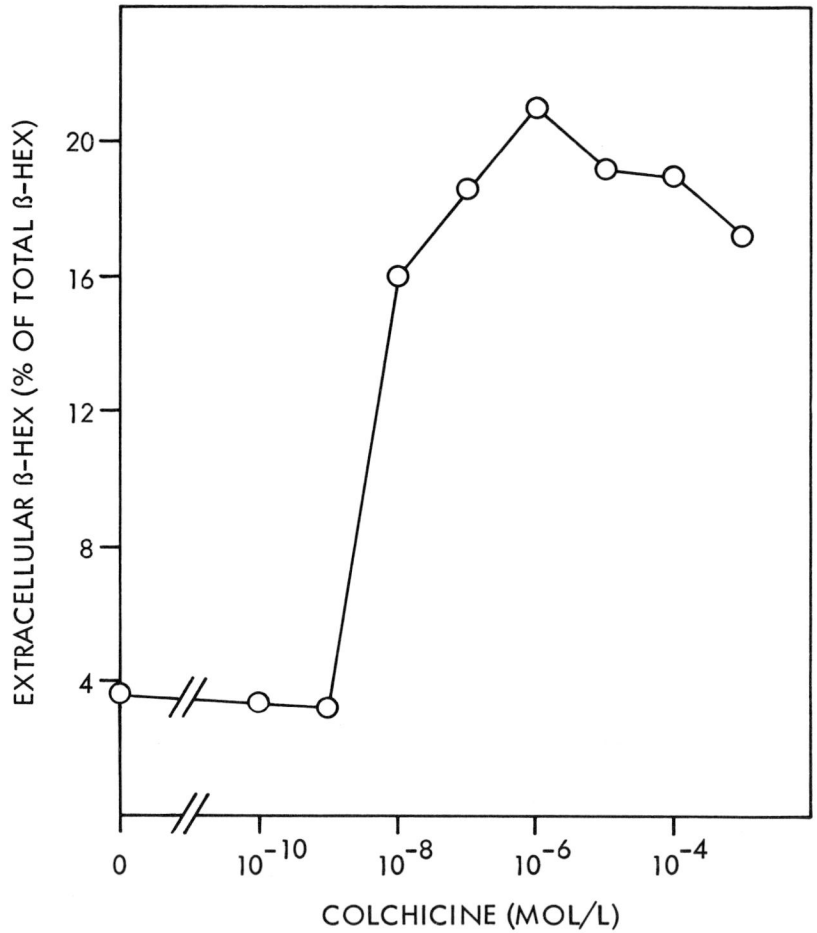

Figure 9.3 Effect of colchicine on the secretion of β-hexosaminidase. Twenty-four hours after feeding the drug was added and the enzyme release was measured during the subsequent 48 h

into the cell the hydrolase reaches the lysosomes in a catalytically active form. This metabolic correction, which has been studied most extensively in genetic mucopolysaccharidoses, represents an *in vitro* model for enzyme replacement therapy in lysosomal storage diseases.

The mechanism by which lysosomal hydrolases are taken into cells, closely

Table 9.1 **Rate of clearance of lyosomal hydrolases and albumin from the medium by fibroblasts***

	Source	Rate of clearance $(ml \times h^{-1} \times mg^{-1})$
α-N-acetylglucosaminidase	human urine	0.065–0.325
	human liver	0.0045
	human serum	0.0049
	human placenta	0.0035–0.0163
	bovine liver	0.0030
β-glucuronidase	human liver ▲	0.0022
	human spleen O	up to 0.085
α-mannosidase △	pig kidney	0.015
β-N-acetylglucosaminidase ●	human skin fibroblasts (intracellular and secreted enzyme)	0.04(0.01–0.1)
α-L-iduronidase ☐	human urine	up to 0.145
[^{125}I]Albumin	human serum △	0.0006
	bovine plasma▲	0.0004

* The values have been calculated from own data and the literature. The rate of clearance represents that volume of medium in millilitres cleared of enzyme per milligram cell protein per hour

△ von Figura, Kresse and Mersmann, unpublished; ▲ Lagunoff *et al.*, 1973; O Nicol *et al.*, 1974; ● Hickmann *et al.*, 1974; ☐ Shapiro *et al.*, 1976

resembles the mechanisms for the uptake of other macromolecules. This uptake is an energy dependent process, which requires the engulfment of the cell membrane, the formation of a vesicle containing the extracellular material and the detachment of this vesicle from the membrane. A number of terms have been used to describe this process among which pinocytosis and endocytosis have been used most commonly (Jacques, 1969; Gordon, 1973).

Two forms of pinocytosis can be differentiated according to their kinetics. Fluid pinocytosis designates a process where the concentration of the macromolecule within the pinosome is directly related to its concentration in the medium. Fluid pinocytosis is non-specific and much less efficient than the second form of pinocytosis, the adsorptive or surface pinocytosis. Adsorptive pinocytosis requires specific interaction between the cell membrane and the macromolecule. Cell membrane receptors and recognition sites on the macromolecules mediate this interaction. If the uptake is measured as a function of the extracellular concentration of a macromolecule, a saturation curve can be plotted. The shape of the curve is determined by the affinity of the macromolecule for the cell membrane receptor and the number of cell receptors available for pinocytosis.

A number of acid hydrolases are known to be ingested by cellular adsorptive pinocytosis (Hickman and Neufeld, 1972; von Figura and Kresse, 1974a; Brot *et al.*, 1974; Lagunoff *et al.*, 1973; Nicol *et al.*, 1974; Mersmann *et al.*, 1976). The uptake rates of different hydrolases vary greatly; even for a single

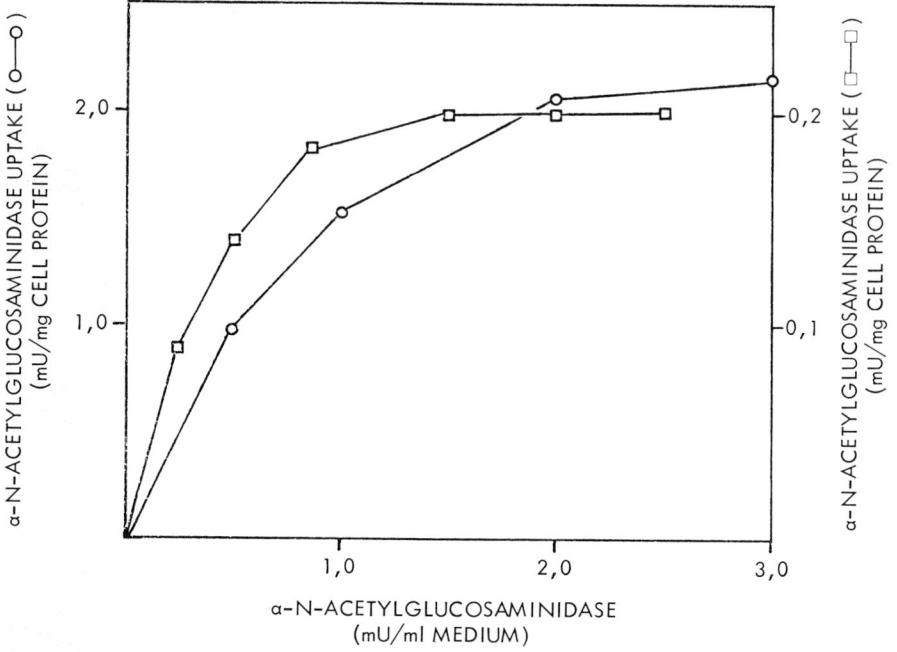

Figure 9.4 Uptake of two different α-N-acetylglucosaminidase preparations from human urine. The period of pinocytosis was 8 h

hydrolase, uptakes can vary when there are differences in the source of that hydrolase (Table 9.1).

Figure 9.4 shows that the uptakes of different preparations of α-N-acetylglucosaminidase from one source may vary greatly. Similar marked differences in uptakes have been reported for β-N-acetylglucosaminidase (Hickman *et al.*, 1974) and β-glucuronidase (Brot *et al.*, 1974; Nicol *et al.*, 1974).

In the course of purification of β-glucuronidase (Nicol *et al.*, 1974; Glaser *et al.*, 1975), α-L-iduronidase (Shapiro *et al.*, 1976) and β-N-acetylglucosaminidase (von Figura, unpublished), it has recently been observed that these enzymes can be fractionated by various techniques into multiple forms. Some so called high-uptake or corrective forms are rapidly taken up by the cells, whereas others, the low-uptake or non-corrective forms are taken into cells at markedly lower rates. The high-uptake forms of β-N-acetylglucosaminidase (Figure 9.5) and β-glucuronidase have a lower pI than the low-uptake forms. The high-uptake form of β-L-iduronidase has a higher molecular weight than the low-uptake form. Catalytic properties of the high and low uptake forms of each of the three enzymes towards synthetic sub-

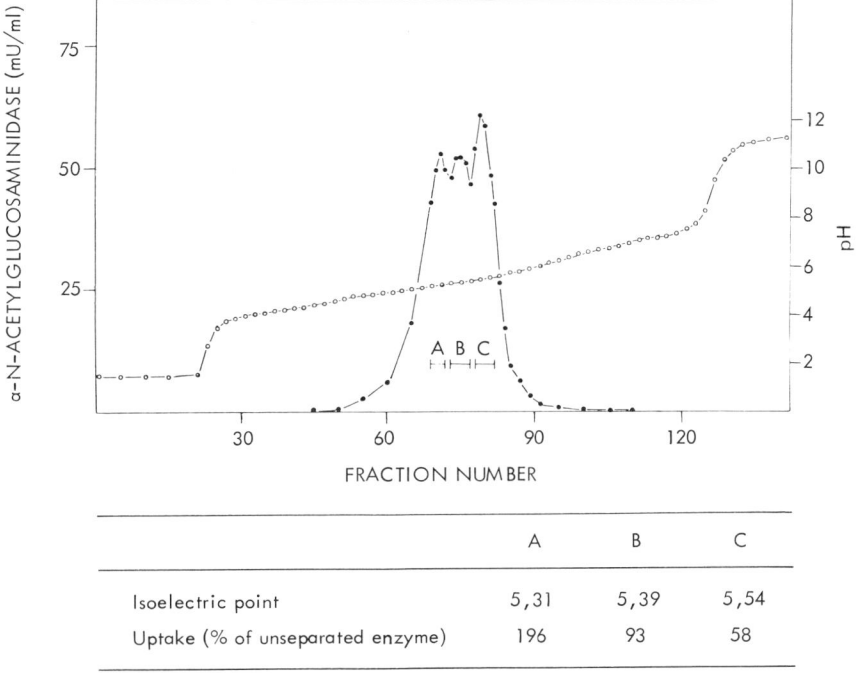

	A	B	C
Isoelectric point	5,31	5,39	5,54
Uptake (% of unseparated enzyme)	196	93	58

Figure 9.5 Isoelectric focusing of partially purified urinary α-N-acetylglucosaminidase (●—●). Isoelectric focusing was performed in a 110 ml column (LKB Produktor AB) according to the producer's manual. A mixture of Ampholine pH 4–6 and Ampholine pH 5–8 (4:1, final concentration 0.7%, v/v) was used to establish the pH gradient (○—○). Fractions corresponding to isoenzymes A, B and C were dialyzed and the relative rates of uptake of isoenzymes A, B and C by fibroblasts were determined

strates do not differ. Different uptake properties of the various enzyme preparations may readily be explained by the variable composition of high and low uptake forms.

Available evidence suggests that there may be structural differences between the high and low uptake forms in that part of the molecule which is responsible for proper recognition of the enzyme by the cell receptor. Periodate oxidation under conditions which do not greatly affect the catalytic activity of β-N-acetylhexosaminidase (Hickman *et al.*, 1974) and α-N-acetylglucosaminidase (Figure 9.6) converts the high uptake forms of these enzymes into low uptake forms. This suggests that the recognition sites of these enzymes consist of carbohydrates. However, all attempts to convert the high uptake forms of β-glucuronidase and α-N-acetylglucosaminidase by treatment with specific glycosidases into low uptake forms have so far failed. If, indeed, carbohydrate structures are involved in the membrane recognition site of the lysosomal enzymes, the structural features required for recognition must differ from those required for the uptake of glycoproteins into hepatocytes (Ashwell and Morrel, 1974). Glycoproteins with exposed terminal β-galactosyl residues, which are

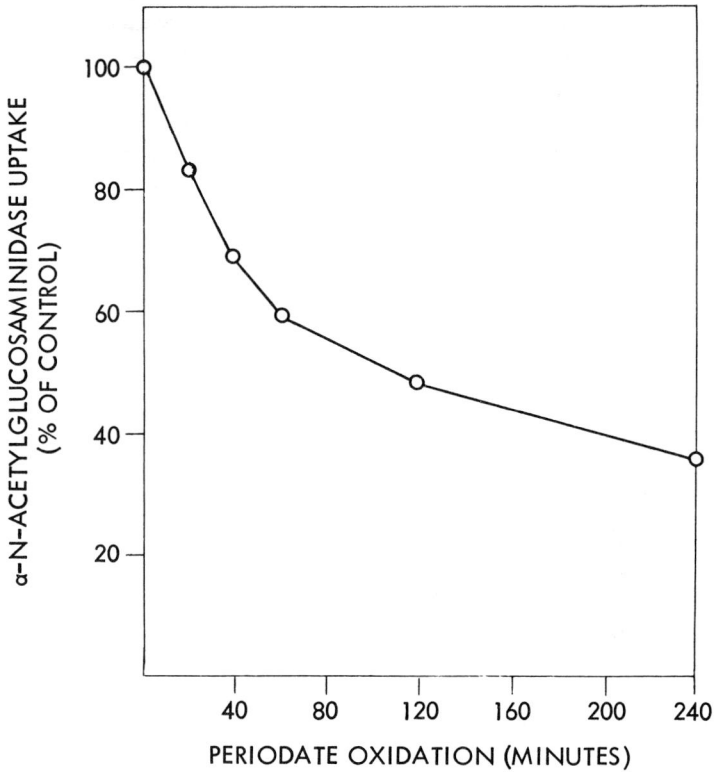

Figure 9.6   Effect of periodate oxidation on the uptake of urinary α-N-acetylglucosaminidase. Equal amounts of enzyme activity recovered after periodate oxidation were then assayed for uptake thus correcting for destruction of activity. The enzyme was treated for specified periods of time with 10 mM periodate at 4 °C in the dark. Periodate oxidation for 240 min destroyed 30% of the α-N-acetylglucosaminidase activity

preferentially taken up into hepatocytes, neither compete with lysosomal hydrolase for uptake (Hickman *et al.*, 1974) nor are themselves taken into fibroblasts (Shapiro *et al.*, 1976).

Preliminary results derived from uptake studies using fibroblasts pretreated with sugar specific lectins (Figure 9.7) suggest that the cell receptor has a carbohydrate structure. The elucidation of the structural requirements for proper recognition of the lysosomal hydrolases should be of practical importance since high uptake forms might be used for enzyme replacement therapy.

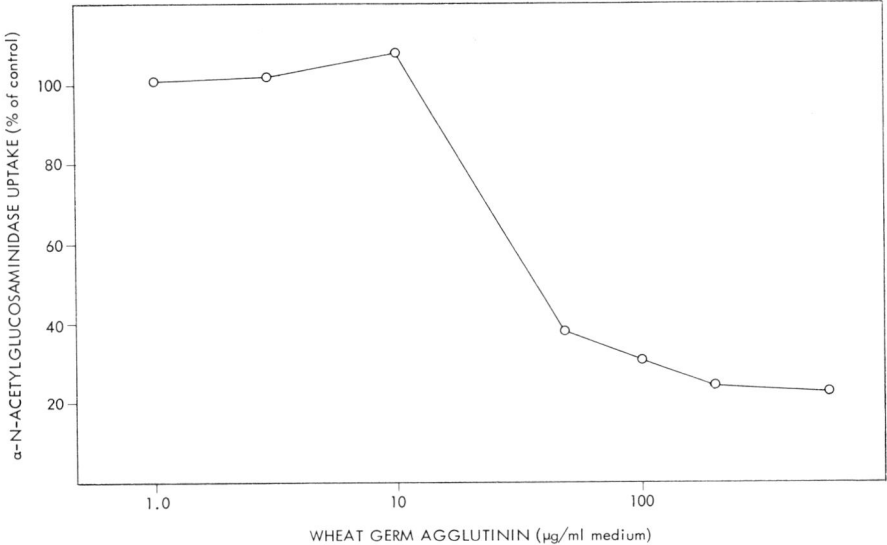

Figure 9.7 Uptake of urinary α-N-acetylglucosaminidase by fibroblasts pretreated with wheat germ agglutinin. Fibroblasts were pretreated with various lectin concentrations for 15 min and uptake of α-N-acetylglucosaminidase was determined after removal of unbound lectin. The lectin effect could be prevented almost completely by the presence of 50 mM N-acetylglucosamine during the preincubation period in the presence of the lectin

Cytochalasin B, colchicine and chloroquine, which stimulate the secretion of lysosomal enzymes, inhibit their uptake. These effects indicate that microfilaments and microtubules participate in the formation of pinosomes and that chloroquine seems to compete with the lysosomal hydrolases for binding to the cell membrane.

THE HICKMAN–NEUFELD CONCEPT FOR THE FORMATION OF SECONDARY LYSOSOMES IN FIBROBLASTS

After their formation, lysosomal hydrolases are packaged into small vesicles. Until recently it was generally accepted that these vesicles (primary lysosomes) would coalesce with pinocytotic or autophagic vesicles containing the substrates of the enzymes (Cohn and Fedorko, 1969). In 1972, Hickman and

Neufeld proposed a new concept which was based on experiments with fibroblasts from patients with I-cell disease (ICD, mucolipidosis II). ICD-fibroblasts exhibit a marked deficiency of a number of acid hydrolases, which in turn are found in excessive quantities in the culture medium of these cells. These extracellular hydrolases are 'internalized' by ICD- and non-ICD-fibroblasts are much lower rates than the corresponding enzymes derived from the secretions of non-ICD-fibroblasts. Hickman and Neufeld concluded from these data, that ICD-hydrolases are defective in the marker required for proper recognition of the enzymes. The particular finding of intracellular deficiency and extracellular accumulation of lysosomal hydrolases under conditions where the uptake of hydrolases is affected, suggests that after their formation lysosomal hydrolases are first secreted and not directly transferred to the secondary lysosomes. To reach the secondary lysosomes the hydrolases have to re-enter the cells by pinocytosis. Since in ICD-fibroblasts some of the lysosomal hydrolases are not affected and others exhibit residual activities of up to 30%, it seems likely that some acid hydrolases are directly transferred to the lysosomes. It may well depend on the particular hydrolase and on the individual cell type, to what extent the enzyme enters the lysosomes directly or via pinocytosis.

Further evidence for the Hickman–Neufeld hypothesis comes from three different experimental approaches:

(1) As mentioned already, cytochalasin B stimulates secretion and inhibits uptake of lysosomal enzymes. Intracellularly the activities of lysosomal enzymes decrease at a rate which is related to their half-lives and storage of sulphated glycosaminoglycans is observed (von Figura and Kresse, 1974a). The storage of substrate and the decrease of enzyme activities at different rates is best explained by the Hickman–Neufeld mechanism, in that inhibition of pinocytosis leads to a half-life dependent decrease of enzyme activities within the secondary lysosomes, which is followed by an impaired degradation of macromolecules such as glycosaminoglycans.

(2) Very recently, Reuser and associates (1976) demonstrated that after prolonged co-cultivation of β-hexosaminidase deficient fibroblasts and normal fibroblasts the intracellular β-hexosaminidase activity decreased in the normal and increased in the deficient fibroblasts. This would again support the Hickman and Neufeld hypothesis. Furthermore, since β-galactosidase and α-glucosidase are not transferred from normal fibroblasts to enzyme deficient fibroblasts it seems likely that the Hickman and Neufeld hypothesis might not apply to all lysosomal enzymes.

(3) Experiments with β-hexosaminidase specific antibodies, which were immobilized by conjugation to Sepharose 4 B and subsequently layered over a culture of normal fibroblasts, showed that the β-hexosaminidase activity bound by the antibodies was about 150% of the β-hexosaminidase activity found in the medium of controls. However, the intracellular β-hexosaminidase activity was not influenced by the immobilization of secreted β-hexosaminidase (von Figura,

unpublished). This suggests that the relative quantity of secreted enzyme, which becomes bound to the antibody–Sepharose, is small compared to that part which enters the lysosome directly or via pinocytosis.

From what is now known about the secretion and the uptake of lysosomal enzymes, it may be suggested that the intracellular enzyme activities of acid hydrolases is not only controlled by their rate of formation and degradation, but also by their secretion and uptake. It remains to be determined to what extent the effects on the intracellular lysosomal enzyme activity of factors such as passage number, cell density, type of pH of medium is mediated through effects on the rate of secretion and uptake.

References

Ashwell, G. and Morrell, A. G. (1974). The role of surface carbohydrates in the hepatic recognition and transport of circulating glycoproteins. *Adv. Enzymol.*, **41**, A. Meister (ed.), p. 99. (New York: Wiley)

Brot, F. E., Glaser, J. H., Roozen, K. J., Sly, W. S. and Stahl, P. D. (1974). *In vitro* correction of deficient human fibroblasts by β-glucuronidase from different human sources. *Biochem. Biophys. Res. Commun.*, **57**, 1

Caro, L. G. and Palade, G. E. (1964). Protein synthesis, storage and discharge in the pancreatic exocrine cell. *J. Cell Biol.*, **20**, 473

Cohn, L. A. and Fedorko, M. E. (1969). The formation and fate of lysosomes. *In*: J. T. Dingle and H. A. Fell (eds.) *Lysosomes in Biology and Pathology*, Vol. 1, pp. 43–63 (Amsterdam: North Holland Publishing Co.)

Dehm, P. and Prockop, J. D. (1972). Time lag in the secretion of collagen by matrix-free tendon cells and inhibition of the secretory process by colchicine and vinblastine. *Biochim. Biophys. Acta*, **264**, 375

Dingle, J. T. (1969). The extracellular secretion of lysosomal enzymes. *In*: J. T. Dingle and H. B. Hall (eds.) *Lysosomes in Biology and Pathology*, Vol. 2, pp. 421–436. (Amsterdam: North Holland Publishing Co.)

de Duve, C. and Wattiaux, R. (1966). Functions of lysosomes. *Annu. Rev. Physiol.*, **28**, 435

Ericson, J. L. E. (1969). Mechanism of cellular authophogy. *In*: J. T. Dingle and H. B. Fell (eds.) *Lysosomes in Biology and Pathology*, Vol. 2, pp. 345–394. (Amsterdam: North Holland Publishing Co.)

Estensen, R. D. and Plagemann, P. G. W. (1972). Cytochalasin B inhibition of glucose and glucosamine transport. *Proc. Nat. Acad. Sci. U.S.A.*, **69**, 1430

von Figura, K. and Kresse, H. (1974a). Quantitative aspects of pinocytosis and the intracellular fate of N-acetyl-α-D-glucosaminidase in Sanfilippo B fibroblasts. *J. Clin. Invest.*, **53**, 85

von Figura, K. and Kresse, H. (1974b). Inhibition of pinocytosis by cytochalasin B. *Eur. J. Biochem.*, **48**, 357

Glaser, J. H., Roozen, K. J., Brot, F. E. and Sly, W. S. (1975). Multiple isoelectric and recognition forms of human β-glucuronidase activity. *Arch. Biochem. Biophys.*, **166**, 536

Goldstone, A. and Koenig, H. (1970). Lysosomal hydrolases as glycoproteins. *Life Sci.*, **9**, 1341

Goldstone, A. and Koenig, H. (1973a). Isolation and characterisation of a rough microsomal fraction. *Biochem. J.*, **132**, 259

Goldstone, A. and Koenig, H. (1973b). Physiochemical modification of lysosomal hydrolases during intracellular transport. *Biochem. J.*, **132**, 267

Gordon, A. H. (1973). The role of lysosomes in protein catabolism. *In*: J. T. Dingle and H. B. Fell (eds.). *Lysosomes in Biology and Pathology*, Vol. 3, pp. 89–137. (Amsterdam: North Holland Publishing Co.)

Hers, H. G. and von Hoof, F. (1973). *Lysosomes and Storage Diseases.* (New York and London: Academic Press)

Hickman, S. and Neufeld, E. F. (1972). A hypothesis for I-cell disease—defective hydrolases that do not enter lysosomes. *Biochem. Biophys. Res. Commun.,* **49,** 992

Hickman, S., Shapiro, L. J. and Neufeld, E. F. (1974). A recognition marker required for uptake of lysosomal enzyme by cultured fibroblasts. *Biochem. Biophys. Res. Commun.,* **57,** 55

Holtzman, E. (1969). Lysosomes in the physiology and pathology of neurons. *In:* J. T. Dingle and H. B. Fell (eds.). *Lysosomes in Biology and Pathology,* Vol. 1, pp. 192–216. (Amsterdam: North Holland Publishing Co.)

Ide, H. and Fishman, W. H. (1969). Dual localisation of beta-glucuronidase and acid phosphatase in lysosomes and in neurosomes. II. Membrane associated enzymes. *Histochemie,* **20,** 300

Jacques, P. J. (1969). Endocytosis. *In:* J. T. Dingle and H. B. Fell (eds.). *Lysosomes in Biology and Pathology,* Vol. 2, pp. 395–420. (Amsterdam: North Holland Publishing Co.)

Jamieson, J. D. and Palade, G. E. (1967a). Intracellular transport of secretory proteins in the pancreatic exocrine cell. I. Role of the peripheral elements in the Golgi complex. *J. Cell Biol.,* **34,** 577

Jamieson, J. D. and Palade, G. E. (1967b). Intracellular transport of secretory proteins in the pancreatic exocrine cell. II. Transport to condensing vacuoles and zymogen granules. *J. Cell Biol.,* **34,** 597

Kato, K., Ide, H, Shirahama, T. and Fishman, W. H. (1970). Incorporation of [14C]glucosamine and [14C]leucine into mouse kidney β-glucuronidase induced by gonadotrophin. *Biochem. J.,* **117,** 161

Kato, K., Hirohata, I., Fishman, W. H. and Tsukamoto, H. (1972). Intracellular transport of mouse kidney β-glucuronidase induced by gonadotrophin. *Biochem. J.,* **127,** 425

Kletzien, R. F., Perdue, J. F. and Springer, A. (1972). Cytochalasin A and B, inhibition of sugar uptake in cultured cells. *J. Biol. Chem.,* **247,** 2964

Koenig, H. (1969). Lysosomes in the nervous system. *In:* J. T. Dingle and H. B. Fell (eds.). *Lysosomes in Biology and Pathology,* Vol. 2, pp. 111–162. (Amsterdam: North Holland Publishing Co.)

Lagunoff, D., Nicol, D. M. and Pritz, P. (1973). Uptake of beta-glucuronidase by deficient human fibroblasts. *Lab. Invest.,* **29,** 449

van Lanckner, J. L. and Lentz, P. L. (1970). Study on the site of biosynthesis of β-glucuronidase and its appearance in lysosome in normal and hypoxic rats. *J. Histochem. Cytochem.,* **18,** 529

Lengsfeld, A. M., Löw, I., Wieland, T., Dancker, P. and Hasselbalch, W. (1974). Interaction of phalloidin with actin. *Proc. Nat. Acad. Sci. U.S.A.,* **71,** 2803

Maunsbach, A. B. (1969). Functions of lysosomes in kidney cells. J. T. Dingle and N. B. Fell (eds.). *In: Lysosomes in Biology and Pathology,* Vol. 1, pp. 115–154. (Amsterdam: North Holland Publishing Co.)

Mersmann, G., von Figura, K. and Buddecke, E. (1967). Mannosidosis. Storage of mannose-containing material in cultured human mannosidosis cells and metabolic correction by pig kidney α-mannosidase. *Hoppe-Seyler's Z. Physiol. Chem.,* **357,** 641

Nichols, B. A., Bainton, D. F. and Farquhar, M. G. (1971). Differentiation of monocytes; origin, nature and date of their azurophil granules. *J. Cell Biol.,* **50,** 498

Nicol, D. M., Lagunoff, D. and Pritzl, P. (1974). Differential uptake of human β-glucuronidase isoenzymes from spleen by deficient fibroblasts. *Biochem. Biophys. Res. Commun.,* **59,** 941

Novikoff, A. B., Essner, E. and Quintana, N. (1964). Golgi apparatus and lysosomes. *Fed. Proc. Fed. Am. Soc. Exp. Biol.,* **23,** 1010

Novikoff, A. B. (1967). Enzyme localization and ultrastructure of neurons. *In:* H. Hyden (ed.), *The Neuron,* pp. 255–318. (New York: American Elsevier Publishing Co.)

Reuser, A., Halley, D., deWit, E., Hogeveen, A., van der Kamp, M., Mulder, M. and Galjaard, H. (1967). Intercellular exchange of lysosomal enzymes: enzyme assays in single human fibroblasts after co-cultivation. *Biochem. Biophys. Res. Commun.,* **69,** 311

Shapiro, L. J., Hall, C. W., Leder, I. G. and Neufeld, E. F. (1976). The relationship of α-L-iduronidase and Hurler corrective factor. *Arch. Biochem. Biophys.*, **172**, 156

Walkinshaw, C. H. and van Lanckner, J. L. (1964). Molecular mechanising of liver regeneration. II. Intracellular distribution of biochemical markers in normal and regenerating liver. *Lab. Invest.*, **13**, 513

Wessels, N. K., Spooner, B. S., Ash, J. F., Bradley, M. O., Luduena, M. A., Taylor, E. L., Wrenn, J. T. and Yamada, K. M. (1971). Microfilaments in cellular and developmental processes. *Science*, **171**, 135

Wiesmann, U. N., DiDonato, S. and Herschkowitz, N. N. (1975). Effect of chloroquine on cultured fibroblasts: release of lysosomal hydrolases and inhibition of their uptake. *Biochem. Biophys. Res. Commun.*, **66**, 1338

10
Sulphatase Deficiencies

A. B. ROY

Department of Physical Biochemistry, John Curtin School of Medical Research, Australian National University, Canberra, Australia

Although the existence of sulphatases has been known for over 60 years, and some of these enzymes have been extensively studied for the last 25, it is only relatively recently that their importance in medicine and in biochemistry has been appreciated. Some idea of this importance is given by Table 10.1 which lists those diseases associated with defects in one or more of the sulphatases. To this list might be added inclusion or I-cell disease (mucolipidosis II) in which the sulphatases are also defective, although not inactive as they are in the diseases listed in Table 10.1. In I-cell disease there is a block in the normal

Table 10.1 Sulphatase deficiencies

Metachromatic leucodystrophy (Sulphatidosis)
Maroteaux–Lamy syndrome (Mucopolysaccharidosis type VI)
Hunter syndrome (Mucopolysaccharidosis type II)
Morquio's syndrome (Mucopolysaccharidosis type IV)
Sanfilippo A syndrome (Mucopolysaccharidosis type III)
 Multiple sulphatase deficiency
Placental sulphatase deficiency

incorporation of sulphatases into the lysosomes (Hickman and Neufeld, 1972). This disease is not associated with a specific defect in the sulphatases and is a rather general defect of the lysosomal hydrolases, possibly arising from extensive sialylation of the enzyme proteins (Vlalutiu and Rattazzi, 1976), so that it need not be considered further at this stage but is discussed by von Figura (1977, p. 105 this volume) in terms of production, release and uptake of acid hydrolases.

In the present connection the term 'sulphatase' must be considered to be a general one embracing a number of different enzymes catalysing the hydrolysis of several different types of derivative of sulphuric acid. These enzymes may or may not be closely related: this is a topic which has hardly yet been studied.

Three general types of enzyme must be included. First, the sulphatases themselves which catalyse the hydrolysis of sulphate esters (Reaction 1) and include the arylsulphatases, the steroid sulphatases, the carbohydrate sulphatases and some other more or less well defined enzymes. Second, the sulphamatases, catalysing the hydrolysis of certain sulphamates (Reaction 2) and third, the phosphosulphatases, catalysing the hydrolysis of the sulphatophosophate bond (Reaction 3) of adenylyl sulphate and of 3'-phosphoadenylyl sulphate. The relation of the phosphosulphatases to the other two groups is obscure and it is possible that at least some of these enzymes may be phosphatases. Although such a distinction would be theoretically important, in practice the end result of their action would be the same.

$$R.OSO_3^- + H_2O \rightarrow R.OH\uparrow + H^+ + SO_4^{2-} \qquad (1)$$

$$R.NH.SO_3^- + H_2O \rightarrow R.NH_3^+\uparrow + SO_4^{2-} \qquad (2)$$

$$\overset{O}{\underset{|}{\overset{\uparrow}{-O.P}}}OSO_3^- + H_2O \rightarrow \overset{O}{\underset{|}{\overset{\uparrow}{-O.P}}}OH + H^+ + SO_4^{2-} \qquad (3)$$
$$\underset{O^-}{} \qquad\qquad \underset{O^-}{}$$

Mammalian tissues in general, including human tissues, contain many sulphatases. Just how many is still not known but there must be at least seven and probably a great many more. Of these, two have been obtained as homogeneous proteins and some of the others considerably purified. It is appropriate to consider these enzymes briefly before going on to deal with the diseases associated with their deficiencies.

THE ENZYMES

Sulphatases have been found in the lysosomes, the microsomes and the cytosol obtained by the usual cell fractionation techniques, and it is useful to deal with them in terms of their intracellular localization because of the obvious importance this has in any discussion of their function. Further details of the properties of the sulphatases may be found in recent reviews by Dodgson and Rose (1975) and by Roy (1976).

Lysosomal sulphatases

These are the most studied of the mammalian sulphatases and two of them are the only ones which have been obtained as well-characterized homogeneous proteins. The latter are particularly interesting because they have long been known as arylsulphatases (Nicholls and Roy, 1971a), which had no obvious physiological function, and they have only recently been shown to be sulphatases of quite different types.

The sulphatase A of ox liver is the best known of the lysosomal sulphatases. Sulphatase A is of almost ubiquitous occurrence in mammalian tissues and enough is known of other purified (Stevens *et al.*, 1975; Lee and van Etten, 1975) and partly purified (Balasubramanian and Bachhawat, 1975) examples to justify the acceptance of the ox liver enzyme as typical. It is an acidic glycoprotein with a molecular weight of about 107 000 containing eight carbohydrate chains (terminating in sialic acid residues) per molecule and probably being made up of two subunits. It contains an unusually high proportion of proline—about 93 residues per molecule. Isoenzymes of human sulphatase A have been separated but not yet completely characterized (Stevens *et al.*, 1973; Dubois *et al.*, 1975b; Stevens *et al.*, 1976). Sulphatase A was originally described as an arylsulphatase, capable of hydrolysing the sulphate esters of many phenols, and although this activity provided a simple means of assaying the enzyme it was far from obvious what role it could play *in vivo*. Only with the very important work of Mehl and Jatzkewitz (1964; 1968) did it become clear that sulphatase A was in fact a cerebroside sulphatase, hydrolysing the galactose 3-sulphate residues of cerebroside sulphate. This cerebroside sulphatase activity can be studied *in vitro* under quite unphysiological conditions such as a medium of very low ionic strength (Stinshoff and Jatzkewitz, 1975), or in the presence of a bile salt, for example, sodium taurodeoxycholate, and manganous ions (Jerfy and Roy, 1973). It can also be studied under rather more physiological conditions, in the presence of a protein activator (Mehl and Jatzkewitz, 1964, 1968; Fischer and Jatzkewitz, 1975), which approximate to the situation *in vivo*. The protein activator has been isolated and partly characterized by Fischer and Jatzkewitz (1975). It is an acidic glycoprotein of molecular weight 21 000 localized in the freely soluble fraction of the lysosome (Mraz *et al.*, 1976) and which forms a 1 to 1 complex with cerebroside sulphate (Fischer and Jatzkewitz, 1977). This complex must be presumed to be the physiological substrate of sulphatase A. The nature of the substrate in the absence of the activator, as is commonly the case with assays *in vitro*, is still not clear. In the presence of taurodeoxycholate the sulpholipid exists as a mixed micelle of bile salt and cerebroside sulphate (particle weight about 24 000 containing about four molecules of cerebroside sulphate) but otherwise it must be a micelle, or micelle-like aggregate, of cerebroside sulphate itself. These apparently are quite large, containing upwards of 200 molecules of cerebroside sulphate (Jeffrey and Roy, 1977). Does the enzyme bind the entire micelle or does it abstract a single molecule of cerebroside sulphate? Does it react only with the molecular form of cerebroside sulphate which must be presumed to be in equilibrium with the micelles? It must be noted that all the situations so far studied *in vitro* are rather artificial because what is taken up by lysosomes *in vivo* is presumably not simply cerebroside sulphate but rather a fragment of myelin containing cerebroside sulphate as a relatively minor component of a complex mixture of lipids. As some of the other constituents of myelin inhibit the hydrolysis of cerebroside sulphates (Mehl and Jatzkewitz, 1964; Stinshoff and Jatzkewitz, 1975) it is clear that a very complex situation could exist in the lysosome. On the other hand, if the activator protein specifically binds cerebroside sulphate, so removing it from the myelin, to give the true substrate for the enzyme then the situation is to some extent simplified. Another question, although a less practical one, which requires an answer is why is the inhibition by SO_4^{2-} of the arylsulphatase activity of sulphatase A competitive, while that of the cerebroside sulphatase activity is noncompetitive. Although this difference may appear trivial it does imply a considerable unexplained difference between the mechanisms of the two reactions.

Despite these questions, it is clear that sulphatase A can hydrolyse the galactose 3-sulphate residues of cerebroside sulphate which must be presumed to be its physiological substrate. It can also hydrolyse *in vitro* the galactose 3-sulphate residues of psychosine sulphate, lactosyl sulphatide and seminolipid. Presumably these can also be hydrolysed *in vivo* but this has not been shown. Finally, sulphatase A can hydrolyse ascorbic acid 2-sulphate (Roy, 1975; Fluharty *et al.*, 1976) which appears to be a normal metabolite of ascorbic acid, but the physiological significance of this activity, if any, is obscure and it should be recalled that ascorbic acid 2-sulphate resembles an aryl sulphate perhaps more than it does a carbohydrate sulphate.

Sulphatase B is less well known but again the enzyme from ox liver has been obtained as a homogeneous protein (Farooqui and Roy, 1976) and can be regarded as typical because other sulphatases B have similar properties (Bleszynski and Roy, 1973; Agogbua and Wynn, 1976). One of the problems with sulphatase B is its general occurrence in multiple forms. Two have been isolated from ox liver and seven from ox brain. The origin of these, and of the corresponding enzymes in other tissues, is obscure. They may be tissue-specific, truly individual enzymes whose physiological activities have little to do with the arylsulphatase activities by which they are detected, they may be artifacts produced during their preparation, they may be artifacts produced by degradation *in vivo*, or they may be intermediate stages in the biosynthesis of a single sulphatase B. The different forms of the enzyme seem to differ only in their charge. As far as is known they have identical molecular weights and enzymic properties so that they can for many purposes be regarded as a single enzyme, a basic glycoprotein of molecular weight 56 000 which, like sulphatase A, has a high content of proline, about 44 residues per molecule. Again sulphatase B has long been known as an arylsulphatase and it is only recently that its physiological role has been discovered. O'Brien *et al.* (1974) have shown that it is involved in the hydrolysis of dermatan sulphate, presumably the oligosaccharides derived therefrom. This role can be more simply investigated by using biosynthetic uridinediphospho-*N*-acetyl-galactosamine 4-sulphate as the substrate (Fluharty *et al.*, 1975). So far no detailed studies have been made of this aspect of the activity of sulphatase B and its relationship to the activity of the chondrosulphatase, mentioned below, remains to be established. The possibility that this chondrosulphatase is identical to sulphatase B seems high.

It should be noted that Koenig has long claimed a relationship between sulphatases A and B. He first suggested that sulphatase B was produced by the removal of sialic acid residues from sulphatase A by treatment of the latter with neuraminidase (Goldstone et al., 1971) but this was disproved by studies of the isolated enzymes (Graham and Roy, 1973). More recently it has been claimed that an anionic sulphatase (sulphatase A?) is formed from a cationic sulphatase (sulphatase B?) during movement of the latter through the endoplasmic reticulum and Golgi apparatus to the lysosomes (Goldstone and Koenig, 1973) and that sulphatase B can be reformed by intracellular autolysis (Goldstone and Koenig, 1974; Sanghavi and Koenig, 1976). The evidence is not entirely convincing in the absence of actual separation of the enzymes because changes in electrophoretic mobility of lysosomal enzymes can be caused by their association with other substances, for example with chondroitin sulphate (Kint et al., 1973). A final decision must await more detailed structural studies of the isolated sulphatases. It is perhaps pertinent that there is no immunological cross-reaction between sulphatases A and B (Shapira and Nadler, 1975; Rhodes et al., 1976).

Although sulphatases A and B are the most studied of the lysosomal sulphatases, other such enzymes include a chondrosulphatase attacking oligosaccharides from chondroitin 4-sulphate (Tudball and Davidson, 1969), A heparin sulphamatase hydrolysing intact heparin (Friedman and Arsenis, 1974) and the phosphosulphatase, adenylyl sulphate sulphohydrolase (Bailey-Wood et al., 1970). Work on sulphatase deficiencies has shown the existence in the lysosomes of a sulphoiduronate sulphatase (Bach et al., 1973) a heparan N-sulphate sulphatase (a sulphamatase) (Kresse and Neufeld, 1972) and perhaps an enzyme hydrolysing N-acetylgalactosamine 6-sulphate residues (Matalon et al., 1974). The relation of the last two enzymes to heparin sulphamatase and chondrosulphatase respectively remains to be established.

Microsomal sulphatases

The situation in the microsomes is more complex because the enzymes are tightly bound to the membrane system and so are effectively insoluble. There are three major sulphatase activities to be considered: sulphatase C which is an arylsulphatase apparently quite distinct from sulphatases A and B, steroid sulphatase which hydrolyses the 3β-sulphates of 5α and Δ^5 steroids and oestrone sulphatase which hydrolyses oestrone sulphate and, presumably, other oestrogen sulphates. The existence of these activities is clearly established but what is not established is how many separate enzymes are involved. In the past it had been considered that all three activities were due to separate enzymes but it now seems likely that the sulphatase C and oestrone sulphatase activities are due to a single protein (Dolly *et al.*, 1972; Iwamori *et al.*, 1976). What relationship the steroid sulphatase activity has to sulphatase C and oestrone sulphatase activities is not yet clear. Iwamori and co-workers (1976) suggested that it is associated with a distinct protein but the evidence is not convincing and this view cannot be regarded as proven. Study of these relatively insoluble microsomal enzymes is difficult because most methods of 'solubilization' have given 'soluble' enzymes in poor yield and in a quite uncharacterized form. Recent work (Iwamori *et al.*, 1976) seems particularly important because of the claim that these enzymes were rather efficiently solubilized by using the dipolar surfactant Mironal H2M. The products were claimed to remain soluble after removal of the surfactant but more detailed studies of this particular point are required.

Microsomes may also contain phosphosulphatase but little useful information is available about this activity.

Sulphatases in the cytosol

As far as is known at present, the only sulphatases in the cytosol are an adenylyl sulphate sulphohydrolase (Stokes *et al.*, 1973a, 1973b) and a 3'-phosphoadenylyl sulphate sulphohydrolase (Denner *et al.*, 1973). The former has been obtained as a homogeneous protein and to some extent characterized.

THE DISEASES

Metachromatic leucodystrophy

This is by far the most studied of the sulphatase deficiencies. The demonstration that the disease was associated with a deficiency of sulphatase A (Austin *et al.*, 1964), alias cerebroside sulphatase (Mehl and Jatzkewitz, 1964), came just as a homogeneous preparation of sulphatase A became available (Nichol and Roy, 1964) so that fairly rapid progress was possible. There is now unambiguous evidence that metachromatic leucodystrophy is due to a specific deficiency of sulphatase A or cerebroside sulphatase. What is not clear is why there is a number of different forms of the disease, differing in their age of onset (Moser, 1972). Nor is it clear if the changes in metachromatic leucodystrophy can be considered as being due entirely to the accumulation of cerebroside sulphate and lactosyl sulphatide (Harzer and Benz, 1974) with the concomitant changes in, and destruction of, the myelin. As has already been pointed out, sulphatase A can also hydrolyse other substances containing galactosyl 3-sulphate residues such as psychosine sulphate and seminolipid which might be expected to accumulate in metachromatic leucodystrophy. This has not been established. Psychosine sulphate is not known to occur naturally (Eto *et al.*, 1974c) and seminolipid is apparently not increased in the testes in metachromatic leucodystrophy (Yamaguchi *et al.*, 1975). The latter finding may reflect no more than the fact that seminolipid does not normally appear in human testes before the age of ten years (Ishizuka et al., 1976). Recently it has been suggested (Fluharty et al., 1976) that an accumulation of ascorbic acid 2-sulphate might occur in metachromatic leucodystrophy and that this might cause changes in the central nervous system. There is no direct evidence for this suggestion which is based simply on the fact that the brain contains more ascorbic acid than it does glucose and on the hypothesis that ascorbic acid 2-sulphate, a substrate for sulphatase A, is a significant metabolite of ascorbic acid. Nevertheless, further investigation of this possibility would be worthwhile because ascorbic acid sulphate is, under oxidizing conditions in vitro, a rather powerful sulphating agent and although there is no convincing evidence that such reactions can occur in vivo there is no reason why they should not. If they could occur, then an accumulation of ascorbic acid 2-sulphate might have very far-reaching effects.

From an enzymological point of view, the main interest in metachromatic leucodystrophy must focus on the reason for the lack of sulphatase A activity. As might have been expected, this has been shown to be due to the production of normal amounts of a protein chemically similar to sulphatase A but with a greatly decreased enzyme activity. The protein is immunologically similar (Stumpf et al., 1971) but not identical (Neuwelt et al., 1973) to normal sulphatase A. A question which has evoked some interest is whether or not the altered sulphatase A produced in metachromatic leucodystrophy is completely devoid of enzyme activity. This is a difficult question to answer because any activity is slight and it is hard to exclude the possibility that it is due to the presence in the concentrated preparations of a contaminating sulphatase of some other type, not necessarily sulphatase B upon which most attention has been directed. In order to determine whether trace amounts of 'sulphatase A' might remain in metachromatic leucodystrophy, long times of incubation are required and the interpretation of such assays is fraught with danger (Nicholls and Roy, 1971b). Nevertheless, Stumpf and Austin (1971) claim that there is a residual sulphatase A activity in metachromatic leucodystrophy and they further suggested that there are differences between the kinetic parameters of the normal enzyme and the enzymes present in each of the three forms of the disease. Their evidence is not entirely convincing, and in fact Shapira and Nadler (1975) state, from immunological studies, that the residual activity is due to traces of sulphatase B. Certainly, from a purely practical point of view, sulphatase A can certainly be considered as completely lacking in metachromatic leucodystrophy. It does, however, seem unwise to assume that in metachromatic leucodystrophy, or indeed in any enzyme defect, there can be only one type of defective enzyme. It must be presumed that there are many possible amino acid substitutions each of which would give a protein having an altered enzyme activity. Until disproved, it would appear necessary to assume that the defective enzymes from different families affected with metachromatic leucodystrophy might differ from one another. This must, of course, complicate detailed chemical studies of these proteins because it would be most unwise to combine preparations from different families. It also follows that there could be a residual sulphatase A activity in some cases but not in others.

Another point must be made. Cultured fibroblasts from patients with metachromatic leucodystrophy can be 'cured' by growing them in a medium containing sulphatase A which is taken up by the cells and incorporated into their lysosomes. The enzyme so taken up is apparently considerably more stable than that in the medium. In the cells, sulphatase A is stable for a week or more whereas that in the culture fluid has a half-life of less than a day (Porter *et al.*, 1971). No explanation of the increased stability has been offered, but due attention does not seem to have been paid to the fact that sulphatases A are, in general, quite unstable in dilute solution (Stevens *et al.*, 1975; Jerfy *et al.*, 1976) and the increased stability *in vivo* may be a function only of the enzyme concentration. The following calculations are based on published data for the specific activity of human sulphatase A (Stevens *et al.*, 1975). In culture fluid (Porter *et al.*, 1971) the concentration of sulphatase A was about 0.6 μg/ml, at which concentration it is quite unstable (Stevens *et al.*, 1975), although some stabilization might be brought about by proteins in the culture fluid. In the 'cured' fibroblasts the activity was 0.4 unit/mg protein or about 0.01 μg sulphatase/mg fibroblast. If the lysosomal volume in the fibroblast is assumed to be similar to that in rat liver, 4 μl/g (Beaufay, 1969), this means that the concentration of sulphatase A in the lysosomes is about 3 mg/ml, at which concentration it is quite stable. The stability of sulphatase A in the lysosomes compared with that in the culture fluid may therefore be more apparent than real, and reflect only its high concentration in lysosomes. Although a similar value of about 1 mg/ml can be derived for the concentration of sulphatase A in ox liver lysosomes, too much emphasis must not be put on the figures. However, it is clear that fibroblasts can concentrate sulphatase A by a factor of 10^3–10^4. This is based on the assumption that all the sulphatase A has been incorporated into the lysosomes. Concentrations in the cell as a whole are lower than that in the medium; a finding confirmed by O'Brien and colleagues (1973) in their investigation of the uptake of α-acetylglucosaminidase by cultures of Sanfilippo B fibroblasts. The mechanism of uptake is unclear. Bach and co-workers (1972) concluded that the uptake of α-L-iduronidase by Hurler fibroblasts was selective because up to 40% of the available enzyme, compared with about 1% of serum albumin, could be taken up by the cells. How this specificity is achieved is a most interesting problem.

Another observation which remains unexplained is the finding of an apparently normal male, 48 years of age, whose peripheral leucocytes have a level of sulphatase A activity only about 10% of the normal heterozygote level (Dubois *et al.*, 1975a). Most of the eight children in his family have similarly low levels of sulphatase A in peripheral leucocytes but only three of them have been shown to have metachromatic leucodystrophy. The explanation of this situation, which has since been found in other families (Lott *et al.*, 1976), is not obvious. Admittedly Sanfilippo B fibroblasts are 'cured' by a relatively small uptake of α-acetylglucosaminidase—2–5% of the normal intracellular level gives a 40–70% 'cure' (O'Brien et al., 1973)—but if 10% of the normal level of sulphatase A is sufficient for the normal metabolism of cerebroside sulphate in some cases, then why is it not sufficient in all cases. It may, of course, be that the level of sulphatase A in short-lived peripheral leucocytes is not a true measure of the level in tissues in which cells often have longer lives (Harkness, 1977, p. 90 this volume), but so far there is no evidence pointing to this.

Little is known of the genetic control of sulphatase A in mammalian cells. No purely genetic studies have yet been carried out with this enzyme although there has been a recent investigation of the level of sulphatase B in different mouse strains (Daniel, 1976). Cell-fusion studies with human fibroblasts (Hors-Cayla et al., 1974) seem to show that a negative control mechanism, of unknown origin, operates in the case of sulphatase A. Fusion of heteroploid fibroblasts, line D98AH$_2$ with a low sulphatase A activity, with either normal fibroblasts or Hunter fibroblasts, both having a high sulphatase A activity, gave cells with an enzyme activity little higher than that of the heteroploid line. It is of interest that steroid sulphatase, a microsomal enzyme, is also apparently under negative control (McMorris et al., 1974).

Maroteaux–Lamy syndrome

The demonstration that this disorder is associated with a deficiency of sulphatase B is fairly recent. Two groups (Stumpf et al., 1973; Fluharty et al., 1974) showed that tissues or cultured cells from affected individuals had a specific deficiency of sulphatase B, the activity being about 10% of the normal. O'Brien and colleagues (1974) showed that such cultured fibroblasts were deficient in N-acetylgalactosamine 4-sulphatase activity and were unable to desulphate the terminal N-acetylgalactosamine 4-sulphate residues of dermatan sulphate. The implication that sulphatase B and N-acetylgalactosamine 4-sulphatase were identical was obvious, and received further support from the observation that crude preparations of sulphatase B from human placenta could hydrolyse UDP-N-acetylgalactosamine 4-sulphate (Fluharty et al., 1975). Their identity has been made virtually certain by the observation (Farooqui and Roy, 1976) that pure sulphatases B1α and B1β from ox liver can also hydrolyse this substrate. It can therefore be taken that the Maroteaux–Lamy syndrome is caused by a deficiency of sulphatase B, or N-acetylgalactosamine 4-sulphatase, which leads to an accumulation, and excretion, of dermatan sulphate.

Again the loss of enzymic activity is due to the synthesis, in normal amounts, of a protein immunologically similar to sulphatase B but with, at least in the single case studied, only about 15% of the normal enzymic activity (Shapira et al., 1975).

Hunter Syndrome

The defect in this X-linked mucopolysaccharidosis is in the metabolism of dermatan and heparan sulphates which accumulate in the tissues and are excreted in the urine. They also accumulate in cultured fibroblasts which can be 'cured' by the Hunter factor present in normal cells and urine. This factor has been shown to be a sulphoiduronate sulphatase (Bach *et al.*, 1973; Sjoberg *et al.*, 1973) which is distinct from any of the known sulphatases. Its assay is difficult, but descriptions have been given of methods measuring the hydrolysis of synthetic O-(α-L-idopyronosyluronic acid 2-sulphate)-(1→4)-2,5-anhydro-D-[^{3}H-1] mannitol 6-sulphate (Lim *et al.*, 1974; Liebaers and Neufeld, 1976). Pure sulphatases B1α and B1β from ox liver showed only a very slight sulphoiduronate sulphatase activity when assayed by this method (Farooqui and Roy, 1976) and it seems likely that this must have been due to traces of a contaminating enzyme. A likely contaminant might be one or more of the minor sulphatases B found, for example, in ox brain (Bleszynski and Roy, 1973) and human fibroblasts (Stevens, 1974). These enzymes have not yet been tested for sulphoiduronate sulphatase activity, or indeed for any of the more recently discovered sulphatase activities.

The Hunter syndrome is the only sulphatase deficiency for which an approach to a useful replacement therapy has been developed. The missing enzyme was supplied either by a skin graft (Dean *et al.*, 1975) or, better, by the subcutaneous administration of 2.2×10^8 histocompatible normal fibroblasts (Dean *et al.*, 1976). In the former case the graft was rejected after three months but a reduced secretion of urinary mucopolysaccharide was still obvious after 9 months. The Hunter factor could be detected in the patient's urine so presumably it was released by the graft and taken up by the host cells. Similar changes were found after the administration of fibroblasts. Again, the effect was surprisingly long lasting and there was an apparent general improvement in the patient's condition. Obviously it is a matter of considerable theoretical, and practical, interest that such a prolonged effect can be obtained from so few cells. It seems to imply some recycling of the lysosomal hydrolases, coupled perhaps with a specificity of uptake which probably occurs with isolated fibroblasts.

Morquio's syndrome

Little can be said of the biochemistry of this condition but it has been reported (Matalon *et al.*, 1974) that it is associated with a deficiency, in cultured fibroblasts, of N-acetylgalactosamine 6-sulphatase. This was the first indication of the occurrence in animal tissues of sulphatase attacking chondroitin 6-sulphate, or its degradation products, because the so-called chondrosulphatases, like sulphatase B, seem to attack N-acetylgalactosamine 4-sulphate residues. An interesting report (Donelly *et al.*, 1976) suggests that the action of N-acetylgalactosamine 6-sulphatase on N-acetylgalactosamine 6-sulphate gives not the expected N-acetylgalactosamine but 3,6-anhydrogalactosamine which is found in acid hydrolysates of normal urine but not in urine from patients with Morquio's syndrome.

Sanfilippo A syndrome

The Sanfilippo A factor, which can 'cure' cultured fibroblasts, has been shown to be a heparan N-sulphate sulphatase (a sulphamatase, catalysing reaction (2)) hydrolysing the relatively small form of heparan N-sulphate stored in this disease (Kresse and Neufeld, 1972). A deficiency of such an enzyme could obviously explain the accumulation of heparan sulphate characteristic of the condition. Again the enzyme is not identical with any of the better known sulphatases, and certainly neither sulphatase B1α nor B1β showed any such activity (Farooqui and Roy, 1976), but it could be related to, or identical with, the little known heparin sulphamatases of mammalian tissues (Friedman and Arsenis, 1974; Dietrich, 1970).

Multiple sulphatase deficiency

This most intriguing condition was first thought to be an atypical metachromatic leucodystrophy in which not only sulphatase A was lacking but also sulphatases B and C (Austin *et al.*, 1965). However, it appears to be a separate entity, now often known as Austin's disease, with a deficiency of both the lysosomal sulphatases A and B and the microsomal activities sulphatase C, oestrone sulphatase and steroid sulphatase (Austin 1973; Eto *et al.*, 1974a). This multiple deficiency leads to an accumulation of many types of sulphate ester including cerebroside sulphate, polysaccharides resembling dermatan and heparan sulphates, and some steroid sulphates. It should not be forgotten that the associated profound metabolic disturbance may be secondary to the accumulation of mucopolysaccharides because these themselves inhibit (Avila and Convit, 1975), or otherwise alter (Kint *et al.*, 1973), many lysosomal hydrolase, including sulphatases.

The most interesting feature of the condition is the virtual disappearance of so many different sulphatase activities in both the lysosomes and microsomes. Exactly how many microsomal enzymes are involved is still a matter of opinion. There may be only one if all three activities are due to a single protein or three or more if all are due to separate proteins. Moser and colleagues (1972) have suggested that the condition arises from a defect in a single gene but how this gene could control the biosynthesis of up to five enzymes is by no means clear unless they all have a common subunit. There is at present no evidence either for or against the existence of such a subunit, although it should be noted that there has been no report of an antibody to sulphatase A cross-reacting with sulphatase B, or vice versa (Shapira and Nadler, 1975; Rhodes *et al.*,

1976). Certainly, it is not impossible that a common subunit could exist for sulphatases A and B. It is more difficult to visualize the same subunit being involved in the microsomal sulphatases unless these are in reality only lysosomal enzymes in transit from the endoplasmic reticulum to the lysosomes and as such have their properties so modified by their incorporation into a relatively hydrophobic membrane that they can hydrolyse the rather nonpolar steroid sulphates. There is no evidence for such a view and the insolubility of the microsomal enzymes makes its experimental investigation a matter of some difficulty. If there is a subunit common to sulphatases A and B then this cannot be defective in either metachromatic leucodystrophy or the Maroteaux–Lamy syndrome because, if it were defective, both sulphatases would of necessity be lacking in these conditions. This would seem to imply a fundamental difference between the defect in Austin's disease and in these other conditions, as has indeed been suggested (Eto et al., 1974b). It would seem wise to bear in mind the possible occurrence of differently defective enzymes in different families which could mean that comparisons between Austin's disease and, for example metachromatic leucodystrophy, could give ambiguous results.

There are other features of Austin's disease which require investigation, one of the main ones being the state of other sulphatases. These fall into two groups, the sulphoiduronate sulphatase, N-acetylgalactosamine 6-sulphatase and heparan N-sulphate sulphatase which are known to be defective in the Hunter, Morquio and Sanfilippo A syndromes respectively and the various phosphosulphatases which have not yet been shown to be involved in any enzyme deficiency. If these enzymes are also lacking in Austin's disease then it makes the one-gene hypothesis even more difficult to understand and if they are completely normal in Austin's disease then it would seem to imply that these enzymes form an independent group of sulphatases lacking the common subunit. This would be particularly interesting in the case of the phosphosulphatases which are, as previously noted (Roy, 1971), difficult to differentiate from phosphatases and of which the single purified example does not have the high proline content of sulphatases A and B (Stokes et al., 1973a).

It is therefore of considerable importance that Kresse and von Figura (personal communication) have shown that in multiple sulphatase deficiency both iduronosulphate sulphatase and heparan N-sulphatase are lacking, as had been implied by the work of Eto et al. (1974b) who noted that neither Hunter nor Sanfilippo A fibroblasts, deficient in iduronosulphatase sulphatase and heparan N-sulphatase respectively, were 'cured' by culturing with fibroblasts showing the multiple sulphatase deficiency. This observation brings the total number of deficient enzymes to between five and seven, depending upon the number of different microsomal sulphatases involved. It seems highly probable that all sulphatases, although perhaps not phosphosulphatases (Austin, 1973), will prove to be defective in this condition and it therefore becomes increasingly difficult to believe that all these enzymes contain a common subunit. Some other explanation of the defect must therefore be sought. What this might be is not immediately obvious. An operator or a regulator gene defect is possible although the latter would appear to be excluded by the fact that a protein immunologically similar to sulphatase A is produced in approximately normal amounts (Austin, 1973). The explanation cannot be a simple one because, assuming a similarity between the individual enzyme defects in multiple sulphatase deficiency and defects in the corresponding single-sulphatase deficiencies, at least two chromosomes must be involved. The Hunter syndrome and placental sulphatase deficiency, which involve sulphoiduronate sulphatase and the microsomal sulphatases respectively, are X-linked whereas the other sulphatase deficiencies are autosomal in nature.

There is little doubt that there is much information still to be obtained about this relatively rare condition and there is likewise little doubt that more detailed studies of it are likely to provide information of value not only to the study of sulphatase deficiencies but also to the fundamental biochemistry of these enzymes.

Placental sulphatase deficiency

This is the least understood of the sulphatase deficiencies and is certainly the least important from a practical point of view because its effects are minor. France and Liggins (1969) were the first to note that in certain women showing a greatly decreased urinary oestrogen excretion during pregnancy there was a deficiency of placental steroid sulphatase. Subsequently Fliegner and colleagues (1972) noted a similar deficiency of placental oestrone sulphatase while France and colleagues (1973) showed that the deficiency involved steroid sulphatase (measured with dehydro*epi*androsterone sulphate as substrate), oestrone sulphatase (with oestrone sulphate) and arylsulphatase (with nitrocatechol sulphate) activities. Unfortunately the arylsulphatase was determined under conditions in which it was impossible to differentiate between sulphatases A, B and C. Sulphatase C must be involved because of its known association with at least oestrone sulphatase, and in view of the very severe effects of the known deficiencies of sulphatases A and B it is tempting to suggest that only microsomal sulphatases are deficient in placental sulphatase deficiency. On the other hand, it could be argued that a deficiency of sulphatase A or B does not usually manifest itself until a year or two after birth so that their deficiency in a short-lived organ such as the placenta might be unimportant. Doubts about the true nature of the condition must, however, persist until specific assays for sulphatases A and B (Baum *et al.*, 1959; Hook *et al.*, 1973) and for sulphatase C (Milsom *et al.*, 1972) have been carried out.

Whatever the exact nature of placental sulphatase deficiency its benign nature, at this time, cannot be disputed. Apart from the low oestrogen excretion the pregnancy proceeds normally. France and Downey (1974) have reported that the children from such pregnancies, all males, were normal at an age of 5 years and that they showed no sulphatase deficiency. Unfortunately their evidence for the latter is by no means conclusive because they measured only serum sulphatase which at best is a mixture of sulphatases A and B and could give no information about sulphatase C or the steroid sulphatases. Only specific assays of the different sulphatases in a number of tissues in the child and mother could give unambiguous information.

In two recently reported cases of placental sulphatase deficiency (Jobsis *et al.*, 1976), for which the name X-chromosomal trophoblast sulphatase deficiency has been suggested, the children had skin lesions and a relationship between the sulphatase deficiency, and X-chromosomal ichthyosis has been proposed. It is to be hoped that further information will soon be forthcoming on this relationship and on the statement (Jobsis *et al.*, 1976) that the only placental arylsulphatase involved is sulphatase C and that sulphatases A and B are normal.

SOME FUTURE PROBLEMS

Apart from the major problem of correcting sulphatase deficiencies, or at least alleviating their effects, there are many more biochemical or enzymological problems which require attention. The answers to these would have ramifications spreading far outside the field of sulphatases.

One of these problems is the question of the significance of the quite ubiquitous distribution of sulphatases A, B and C and so, presumably, of the other sulphatases. For example, with sulphatase A it appears that the physiological role of this enzyme is the hydrolysis of cerebroside sulphate, a constituent of myelin and a major component only of the central nervous system. Why, then, are the liver and kidney major sources of sulphatase A? Is this simply a reflection of a general constancy of composition of the lysosomes, or does it mean that sulphatase A and other similar enzymes have functions as yet undiscovered? Perhaps the sulphatase A of liver lysosomes can be transported to the lysosomes of brain where it can exert its physiological role. The long-lasting effects of skin grafts or fibroblast injections in alleviating the sulphoiduronate sulphatase deficiency of the Hunter syndrome (Dean *et al.*, 1975, 1976) support such a view and point to an inter-organ circulation of lysosomal enzymes. If such be the case the rather widespread effects of I-cell disease, in which the uptake of lysosomal hydrolases is defective (Hickman and Neufeld, 1972), become more readily understandable but the major problem posed then becomes the mechanism of the uptake of these enzymes.

The most interesting chemical problem is the identification of the, at present hypothetical, subunit common to sulphatases A, B and C, and perhaps to other sulphatases. Certainly for sulphatases A and B the problem is potentially soluble. If such a subunit exists then there are interesting implications for the evolution of the sulphatases. If it does not exist then there are obvious, and important, problems posed by multiple sulphatase deficiency.

So far no deficiency of a phosphosulphatase has been found. This may reflect no more than the complicated interrelationships of the nucleotidases and sulphatases which participate in the metabolism of the sulphatophosphates (Dodgson and Rose, 1975) and make it difficult to determine the individual enzymes involved. On the other hand, it may be that a deficiency of a phosphosulphatase would be incompatible with life. This seems improbable, but it must be admitted that at present no physiological role can be envisaged for the phosphosulphatases. Then, neither were functions for sulphatases A and B envisaged until the appropriate deficiency diseases were discovered. The discovery of a deficiency of a phosphosulphatase might well shed some light on these most enigmatic of the sulphatases.

References

Agogbua, S. I. O. and Wynn, C. H. (1976). Purification and properties of arylsulphatase B of human liver. *Biochem. J.*, **153,** 415

Austin, J. H. (1973). Multiple sulfatase deficiency. *Arch. Neurol., Chicago*, **28,** 258

Austin, J., McAfee, D., Armstrong, D., O'Rourke, M., Shearer, L. and Bachhawat, B. (1964). Abnormal sulphatase activities in two human diseases (metachromatic leucodystrophy and gargoylism). *Biochem. J.*, **93,** 15c

Austin, J. H., Armstrong, D. and Shearer, L. (1965). Metachromatic form of diffuse cerebral sclerosis. V. The nature and significance of low sulfatase activity: a controlled study of brain, liver and kidney in four patients with metachromatic leukodystrophy. *Arch. Neurol. (Chicago)*, **13,** 593

Avila, J. L. and Convit, J. (1975). Inhibition of leucocytic lysosomal enzymes by glycosaminoglycans *in vitro*. *Biochem. J.*, **152,** 57

Bach, G., Friedman, R., Weissman, B. and Neufeld, E. F. (1972). The defect in the Hurler and Scheie syndromes: deficiency of α-L-iduronidase. *Proc. Nat. Acad. Sci. U.S.A.*, **69,** 2048

Bach, G., Eisenberg, F., Cantz, N. and Neufeld, E. F. (1973). The defect in the Hunter syndrome: deficiency of sulphoiduronate sulphatase. *Proc. Nat. Acad. Sci. U.S.A.*, **70,** 2134

Bailey-Wood, R., Dodgson, K. S. and Rose, G. A. (1970). Purification and properties of two adenosine-5'-phosphosulphate sulphohydrolases from rat liver and their possible role in the degradation of 3'-phosphoadenosine-5'-phosphosulphate. *Biochem. Biophys. Acta*, **220,** 284

Balasubramanian, K. A. and Bachhawat, B. K. (1975). Purification, properties and glycoprotein nature of arylsulfatase A from sheep brain. *Biochim. Biophys. Acta*, **403,** 113

Baum, H., Dodgson, K. S. and Spencer, B. (1959). The assay of arylsulphatases A and B in human urine. *Clin. Chim. Acta*, **4,** 453

Beaufay, H. (1969). Methods for isolation of lysosomes. *In*: J. T. Dingle and H. B. Fell (eds.). *Lysosomes in Biology and Pathology*, Vol. 2, pp. 515–546. (Amsterdam: North-Holland Publishing Co.)

Bleszynski, W. S. and Roy, A. B. (1973). Some properties of the sulphatase B of ox liver. *Biochim. Biophys. Acta*, **317,** 164

Daniel, W. L. (1976). Genetics of murine liver and kidney arylsulfatase B. *Genetics*, **82,** 477

Dean, M. F., Muir, H., Benson, V. F., Button, L. R., Batchelor, J. R. and Bewick, M. (1975). Increased breakdown of glycosaminoglycans and appearance of corrective enzyme after skin transplants in Hunter syndrome. *Nature (London)*, **257,** 609

Dean, M. F., Muir, H., Benson, P. F., Button, L. R., Boylston, A. and Mowbray, J. (1976). Enzyme replacement therapy by fibroblast transplantation in a case of Hunter syndrome. *Nature (London)*, **261,** 323

Denner, W. H. B., Stokes, A. M., Rose, F. A. and Dodgson, K. S. (1973). Separation and properties of the soluble 3'-phosphoadenosine-5'-phosphosulphate-degrading enzymes of bovine liver. *Biochim. Biophys. Acta*, **315**, 394

Dietrich, C. P. (1970). A heparin sulfamidase from mammalian tissues. *Can. J. Biochem.*, **48**, 725

Dodgson, K. S. and Rose, F. A. (1975). Sulfohydrolases. In: D. M. Greenberg (ed.). *Metabolic Pathways*, 3rd Ed., Vol. VII, pp. 359–431. (New York and London: Academic Press)

Dolly, J. O., Dodgson, K. S. and Rose, F. A. (1972). Studies on the oestrogen sulphatase and arylsulphatase C activities of rat liver. *Biochem. J.*, **128**, 337

Donelly, K. A., Di Ferrante, N., Thenet, J.-P. and Dell, J. C. (1976). Occurrence of 3,6-anhydrogalactosamine in human urine. *Abstr. 10th Int. Congr. Biochem.*, p. 596

Dubois, G., Turpin, J.-C. and Baumann, N. (1975a). Absence of arylsulfatase A activity in healthy father of a patient with metachromatic leucodystrophy. *N. Engl. J. Med.*, **293**, 302

Dubois, G., Turpin, J. C. and Baumann, N. (1975b). Arylsulphatases in metachromatic leucodystrophy: detection of a new variant by electrophoresis. Improvement of quantitative assay. *Biomed. Express*, **23**, 116

Eto, Y., Rampini, S., Wiesmann, U. and Herschkowitz, N. N. (1974a). Enzymic studies of sulphatases in tissues of the normal human and in metachromatic leucodystrophy with multiple sulphatase deficiencies: arylsulphatases A, B and C, cerebroside sulphatase, psychosine sulphatase and steroid sulphatases. *J. Neurochem.*, **23**, 1161

Eto, Y., Wiesmann, U. N., Carson, J. H. and Hershkowitz, N. N. (1974b). Multiple sulfatase deficiencies in cultured skin fibroblasts. *Arch. Neurol.*, **30**, 153

Eto, Y., Wiesmann, U. and Herschlowitz, N. N. (1974c). Sulfogalactosylsphinogosine sulfatase. Characteristics of the enzyme and its deficiency in metachromatic leucodystrophy in human cultured skin fibroblasts. *J. Biol. Chem.*, **249**, 4955

Farooqui, A. A. and Roy, A. B. (1976). The sulphatase of ox liver. XX. The preparation of sulphatases B1α and B1β. *Biochim. Biophys. Acta*, **452**, 341

Fischer, G. and Jatzkewitz, H. (1975). The activator of cerebroside sulphatase. Purification from human liver and identification as a protein. *Z. Physiol. Chemie*, **356**, 605

Fischer, G. and Jatzkewitz, H. (1977). The activator of cerebroside sulphatase. Binding studies with enzyme and substrate demonstrating the detergent function of activator protein. *Biochim. Biophys. Acta*, **481**, 561

Fliegner, J. R. H., Schindler, I. and Brown, J. B. (1972). Low urinary oestriol excretion during pregnancy associated with placental sulphatase deficiency and congenital adrenal hypoplasia. *J. Obstet. Gynaecol. Brit. Commonw.*, **79**, 810

Fluharty, A. L., Stevens, R. L., Sandars, D. L. and Kihara, H. (1974). Arylsulfatase B deficiency in Maroteaux–Lamy syndrome cultured fibroblasts. *Biochem. Biophys. Res. Commun.*, **59**, 455

Fluharty, A. L., Stevens, R. L., Fung, D., Peak, S. and Kihara, H. (1975). Uridine diphospho-N-acetylgalactosamine-4-sulfate sulfohydrolase activity of human arylsulfatase B and its deficiency in the Maroteaux–Lamy syndrome. *Biochem. Biophys. Res. Commun.*, **64**, 955

Fluharty, A. L., Stevens, R. L., Miller, R. T., Shapiro, S. S. and Kihara, H. (1976). Ascorbic acid 2-sulfate sulfohydrolase activity of human arylsulfatase A. *Biochim. Biophys. Acta*, **429**, 508

France, J. T. and Liggins, G. C. (1969). Placental sulfatase deficiency. *J. Clin. Endocrinol. Metab.*, **29**, 138

France, J. T., Seddon, R. J. and Liggins, G. C. (1973). A study of a pregnancy with low estrogen production due to placental sulfatase deficiency. *J. Clin. Endocrinol. Metab.*, **36**, 1

France, J. T. and Downey, J. A. (1974). A study of arylsulphatase activity in children born of pregnancies affected with placental sulfatase deficiency. *Biochem. Med.*, **10**, 167

Friedman, Y. and Arsenis, C. (1974). Studies on the heparin sulphamidase from rat spleen. Intracellular distribution and characterisation of the enzyme. *Biochem. J.*, **139**, 699

Goldstone, A., Konecny, P. and Koenig, H. (1971). Lysosomal hydrolases. Conversion of acidic to basic forms by neuraminidase. *FEBS Lett.*, **13**, 68

Goldstone, A. and Koenig, H. (1973). Physiochemical modifications of lysosomal hydrolases during intracellular transport. *Biochem. J.*, **132**, 267

Goldstone, A. and Koenig, H. (1974). Autolysis of glycoproteins in rat kidney lysosomes *in vitro*. Effects on the isoelectric focussing behaviour of glycoproteins, arylsulphatase and β-glucuronidase. *Biochem. J.*, **141**, 127

Graham, E. R. B. and Roy, A. B. (1973). The sulphatase of ox liver. XVII. Sulphatase A as a glycoprotein. *Biochim. Biophys. Acta*, **329**, 88

Harzer, K. and Benz, H. U. (1974). Deficiency of lactosyl sulfatide sulfatase in metachromatic leucodystrophy. *Z. Physiol. Chemie*, **355**, 744

Hickman, S. and Neufeld, E. F. (1972). A hypothesis for I-cell disease: defective hydrolases that do not enter lysosomes. *Biochem. Biophys. Res. Commun.*, **49**, 992

Hook, G. E. R., Dodgson, K. S., Rose, F. A. and Worwood, H. (1973). Relative distribution of arylsulphatases A and B in rat liver parenchymal and other cells. *Biochem. J.*, **134**, 191

Hors-Cayla, M. C., Heuertz, S., Picard, J. Y. and Frezal, J. (1974). Regulation in man × man somatic cell hybrids: possible negative control of arylsulphatase A. *Ann. Hum. Genet.*, **38**, 57

Ishizuka, I., Inomata, M., Heno, K., Suzuki, M. and Yamakawa, T. (1976). Complex glycolipids of testes and nervous system in relation to ageing of animals. *Abstr. 10th Inter. Congr. Biochem.*, p. 259

Iwamori, M., Moser, H. W. and Kishimoto, Y. (1976). Solubilization and partial purification of steroid sulfatase from rat liver: characterization of estrone sulfatase. *Arch. Biochem. Biophys.*, **174**, 199

Jeffrey, H. J. and Roy, A. B. (1977). Amphiphilic properties of cerebroside sulphate. *Aust. J. Exp. Biol. Med. Sci*, **55**

Jerfy, A. and Roy, A. B. (1973). The sulphatase of ox liver. XVI. A comparison of the arylsulphatase and cerebroside sulphatase activities of sulphatase A. *Biochim. Biophys. Acta*, **293**, 178

Jerfy, A. and Roy, A. B. (1974). The sulphatase of ox liver. XVIII. An essential histidyl residue in sulphatase A. *Biochim. Biophys. Acta*, **371**, 76

Jerfy, A., Roy, A. B. and Tomkins, H. J. (1976). The sulphatase of ox liver. XIX. On the nature of the polymeric forms of sulphatase A present in dilute solutions. *Biochim. Biophys. Acta*, **422**, 335

Jobsis, A. C., van Duuren, C. Y., de Vries, G. P., Koppe, J. G., Rijken, Y., van Kempen, G. M. J. and de Groot, W. P. (1976). Trophoblast sulphatase deficiency associated with X-chromosomal ichthyosis. *Ned. Tijdschr. Geneesk.*, **120**, 1980

Kint, J. A., Dacremont, G., Carton, D., Orye, E. and Hooft, C. (1973). Mucopolysaccharidoses: secondarily induced abnormal distributions of lysosomal isoenzymes. *Science*, **181**, 352

Kresse, H. and Neufeld, E. F. (1972). The Sanfilippo A corrective factor. Purification and mode of action. *J. Biol. Chem.*, **247**, 2164

Lee, G. D. and van Etten, R. L. (1975). Purification and properties of a homogeneous arylsulfatase A from rabbit liver. *Arch. Biochem. Biophys.*, **166**, 280

Liebaers, I. and Neufeld, E. F. (1976). Iduronate sulfatase activity in serum, lymphocytes and fibroblasts—simplified diagnosis of the Hunter syndrome. *Pediatr. Res.*, **10**, 733

Lim, T. W., Leder, I. G., Bach, G. and Neufeld, E. F. (1974). An assay for iduronosulfate sulfatase (Hunter corrective factor). *Carbohydr. Res.*, **37**, 103

Lott, I., Dulaney, J. T., Milunsky, A., Hoefnagel, D. and Moser. (1976). Apparent biochemical homozygosity in two obligatory heterozygotes for metachromatic leucodystrophy. *J. Paediatr.* **89**, 43

Matalon, R., Arbogast, B., Justice, P., Brandt, I. K. and Dorfman, A. (1974). Morquio's syndrome. Deficiency of a chondroitin sulfate *N*-acetylhexosamine sulfate sulfatase. *Biochem. Biophys. Res. Commun.*, **61**, 759

McMorris, F. A., Kolber, A. R., Moore, B. W. and Perumal, A. S. (1974). Expression of the neuron-specific protein, 14-3-2 and steroid sulfatase in neuroblastoma cell hybrids. *J. Cell. Physiol.*, **84**, 473

Mehl, E. and Jatzkewitz, H. (1964). Eine Cerebrosidsulfatase aus Schweineniere. *Z. Physiol. Chemie*, **339**, 260

Mehl, E. and Jatzkewitz, H. (1968). Cerebroside 3-sulfate as a physiological substrate of arylsulfatase A. *Biochim. Biophys. Acta*, **151**, 619

Milsom, D. W., Rose, F. A. and Dodgson, K. S. (1972). The specific assay of arylsulphatase C, a rat liver microsomal marker enzyme. *Biochem. J.*, **128**, 331

Moser, H. W. (1972). Sulfatide lipidosis: metachromatic leucodystrophy. *In*: J. B. Stanbury, J. B. Wyngaarden and D. S. Fredrickson (eds.). *The Metabolic Basis of Inherited Disease*, 3rd Ed., pp. 688–729. (New York: McGraw-Hill)

Moser, H. W., Sugita, M., Harbison, M. D. and Williams, M. (1972). Liver glycolipids, steroid sulfates and steroid sulfatases in a form of metachromatic leukodystrophy associated with multiple sulfatase deficiencies. *In*: B. W. Volk and S. M. Aronson (eds.). *Sphingolipids, Sphingolipidoses and Allied Disorders*, pp. 429–450. (New York: Plenum Publishing Co.)

Mraz, W., Fischer, G. and Jatzkewitz, M. (1976). The activator of cerebroside sulphatase, Lysosomal localisation. *Z. Physiol. Chemie*, **257**, 1181

Neuwelt, E., Kohler, P. F. and Austin, J. (1973). Primary enzyme immunoassay (PEIA): studies of the mutant enzyme in metachromatic leukodystrophy. *Immunochemistry*, **10**, 767

Nichol, L. W. and Roy, A. B. (1964). The sulphatase of ox liver. VIII. The sedimentation of purified sulphatase A. *J. Biochem., Tokyo*, **55**, 643

Nicholls, R. G. and Roy, A. B. (1971a). Arylsulfatases. *In*: P. D. Boyer (ed.). *The Enzymes*, 3rd Ed., Vol. V, pp. 21–41

Nicholls, R. G. and Roy, A. B. (1971b). The sulphatase of ox liver. XV. Changes in the properties of sulphatase A in the presence of substrate. *Biochem. Biophys. Acta*, **242**, 141

O'Brien, J. S., Miller, A. L., Loverde, A. W. and Veath, M. L. (1973). Sanfilippo disease type B: enzyme replacement and metabolic correction in cultured fibroblasts. *Science*, **181**, 753

O'Brien, J. F., Cantz, M. and Spranger, J. (1974). Maroteaux–Lamy disease (mucopolysaccharidosis VI), subtype A: deficiency of a *N*-acetylgalactosamine 4-sulfatase. *Biochem. Biophys. Res. Commun.*, **60**, 1170

Porter, M. T., Fluharty, A. L. and Kihara, H. (1971). Corrections of abnormal cerebroside sulfate metabolism in cultured metachromatic leukodystrophy fibroblasts. *Science*, **172**, 1263

Rhodes, T. L., Stout, R. L. and Edmond, J. (1976). Studies on arylsulfatase B from human liver. *Fed. Proc. Fed. Am. Soc. Exp. Biol.*, **35**, 1632

Roy, A. B. (1971). The hydrolysis of sulfate esters. *In*: P. Boyer (ed.). *The Enzymes*, 3rd Ed., Vol. V, pp. 1–19. (New York and London: Academic Press)

Roy, A. B. (1975). L-Ascorbic acid 2-sulphate. A substrate for mammalian arylsulphatases. *Biochim. Biophys. Acta*, **377**, 356

Roy, A. B. (1976). Sulphatases, lysosomes and disease. *Aust. J. Exp. Biol. Med. Sci.*, **54**, 111

Sanghavi, P. and Koenig, H. (1976). Autophagy-related changes of arylsulphatases A and B in rat liver lysosomes. *Biochem. J.*, **155**, 725

Shapira, E., De Gregorio, D. R., Matalon, R. and Nadler, H. L. (1975). Reduced arylsulfatase B activity of the mutant enzyme protein in Maroteaux–Lamy syndrome. *Biochem. Biophys. Res. Commun.*, **62**, 448

Shapira, E. and Nadler, H. L. (1975). The nature of residual arylsulfatase activity in metachromatic leucodystrophy. *J. Pediatr.*, **86**, 881

Sjoberg, I., Fransson, L. A., Matalon, R. and Dorfman, A. (1973). Hunter's syndrome: a deficiency of L-iduronosulfate sulfatase. *Biochem. Biophys. Res. Commun.*, **54**, 1125

Stevens, R. L. (1974). Minor anionic arylsulfatases in cultured human fibroblasts. *Biochim. Biophys. Acta*, **379**, 249

Stevens, R. L., Hartman, M., Fluharty, A. L. and Kihara, H. (1973). A second form of sulfatase A in human urine. *Biochim. Biophys. Acta*, **302**, 338

Stevens, R. L., Fluharty, A. L., Skokut, M. H. and Kihara, H. (1975). Purification and properties of arylsulfatase A from human urine. *J. Biol. Chem.*, **250**, 2495

Stevens, R. L., Fluharty, A. L., Killgrove, A. R. and Kihara, H. (1976). Microheterogeneity of arylsulfatase A from human tissues on isoelectric focussing. *Biochim. Biophys. Acta*, **445**, 661

Stinshoff, K. and Jatzkewitz, H. (1975). Comparison of the cerebroside sulphatase and arylsulphatase activity of human sulphatase A in the absence of activators. *Biochim. Biophys. Acta*, **377**, 126

Stokes, A. M., Denner, W. H. B., Rose, F. A. and Dodgson, K. S. (1973a). Purification of a soluble adenosine 5'-phosphosulphate sulphohydrolase from bovine liver. *Biochim. Biophys. Acta*, **302**, 64

Stokes, A. M., Denner, W. H. B. and Dodgson, K. S. (1973b). Kinetic properties of the soluble adenosine 5'-phosphosulphate sulphohydrolase from bovine liver. *Biochim. Biophys. Acta*, **315**, 402

Stumpf, D. and Austin, J. (1971). Metachromatic leukodystrophy (MLD). IX. Qualitative and quantitative differences in urinary arylsulfatase A in different forms of MLD. *Arch. Neurol. (Chicago)*, **24**, 117

Stumpf, D., Neuwelt, E., Austin, J. and Kohler, P. (1971). Metachromatic leukodystrophy. X. Immunological studies of the abnormal sulfatase A. *Arch. Neurol. (Chicago)*, **25**, 427

Stumpf, D. A., Austin, J. H., Crocker, A. C. and La France, M. (1973). Mucopolysaccharidosis type VI (Maroteaux–Lamy syndrome). I. Sulfatase B deficiency in tissues. *Am. J. Dis. Child.*, **126**, 747

Tudball, N. and Davidson, E. A. (1969). Isolation of a novel sulphatase from rat liver. *Biochim. Biophys. Acta*, **171**, 113

Vlalutiu, G. D. and Rattazzi, M. C. (1976). Abnormal lysosomal hydrolases excreted by cultured fibroblasts in I-cell disease (Mucolipidosis II). *Biochem. Biophys. Res. Commun.*, **67**, 956

Yamaguchi, S., Aoki, K., Handa, S. and Yamakawa, T. (1975). Deficiency of seminolipid sulphatase activity in brain tissue of metachromatic leucodystrophy. *J. Neurochem.*, **24**, 1087

11
Clinical, Biochemical and Genetic Heterogeneity in Gangliosidoses

H. GALJAARD and A. J. J. REUSER

Department of Cell Biology and Genetics, Medical Faculty, Erasmus University, Rotterdam, The Netherlands

During the last decade the responsible enzymic defect has been elucidated for a considerable number of genetic metabolic diseases. This has refined the diagnosis and contributed to prevention of these disorders and it has offered new perspectives in basic research on biochemical and genetic aspects of the enzymes involved.

For many years the diagnosis of genetic metabolic disease was exclusively based on clinical symptoms and signs, results of urine and serum analyses and pedigree analysis. Most of the syndromes known to the clinician have been defined in this way and usually patients will be detected and classified accordingly. The possibility of determining the genetic enzyme deficiency is changing this pattern. In principle the diagnosis of a patient can now be established before clinical symptoms become apparent. Also a patient will be classified on the basis of the enzyme deficiency detected independently of the clinical picture. Recently it has become clearer that one particular enzyme deficiency can be accompanied by different overall biochemical, pathological and clinical abnormalities. (See Stanbury *et al.*, 1972; McKusick *et al.*, 1972; Kirkman, 1972; McKusick, 1975.)

Analogous to the large spectrum of variants observed in proteins like haemoglobin and glucose-6-phosphate dehydrogenase (Harris, 1976; McKusick, 1975) it is to be expected that a particular enzyme deficiency may be caused by different gene mutations. Differences in the molecular structure of the deficient enzyme might or might not result in different functional properties which in turn could be responsible for the biochemical and clinical heterogeneity seen in patients.

The fact that a considerable number of genetic enzyme deficiencies are also expressed in cultured skin fibroblasts of patients (Nadler, 1972; Nitowsky, 1972; Neufeld, 1974; McKusick, 1975) greatly contributes to the investigation of the genetic and biochemical background of clinical heterogeneity in metabolic disorders. Cultured fibroblasts provide a homogeneous population of cells which can be used for the study of biochemical and immunological properties of normal and defective enzymes. The ability to store and subculture fibroblasts is especially advantageous in cases of rare metabolic disorders which lead to early death of the patient.

Cell fusion of fibroblasts from two different patients with the same genetic enzyme deficiency enables a complementation analysis to be made, which may in turn answer the question whether or not different gene mutations are involved. Such studies have so far been performed for different clinical variants of xeroderma pigmentosum (de Weerd-Kastelein *et al.*, 1972; Kraemer *et al.*, 1975), G_{M2}-gangliosidosis (Thomas *et al.*, 1974; Galjaard *et al.*, 1974a), G_{M1}-gangliosidosis (Galjaard *et al.*, 1975), maple syrup urine disease (Lyons *et al.*, 1973) and for different types of methylmalonicacidaemia (Gravel *et al.*, 1975).

Hybridization of human cells with those from Chinese hamster or mouse has allowed the localization of one or more markers on each of the human chromosomes (Human Gene Mapping Conferences, 1974, 1975) and may help to elucidate the molecular structure of certain enzymes.

It is the purpose of this paper to illustrate the three experimental approaches mentioned and describe our use of them in the investigation of genetic enzyme deficiencies leading to G_{M1}- and G_{M2}-gangliosidosis.

G_{M1}-GANGLIOSIDOSES

Clinical types

The first reported series of patients with this autosomal-recessive metabolic disease comprised an infantile form characterized by retarded psychomotor development from birth, progressive mental and motor deterioration, severe bony deformities, hepatosplenomegaly and death within 1–2 years (O'Brien, 1972; O'Brien *et al.*, 1972). The massive cerebral and visceral accumulation of G_{M1}-ganglioside was found to be due to a deficiency of β-glactosidase (Okada and O'Brien, 1968). Later, it was shown that this enzyme deficiency was also responsible for the accumulation of the mucopolysaccharide, keratansulphate, in visceral organs (Wolfe *et al.*, 1970).

The β-galactosidase deficiency can be demonstrated in cerebral tissue, visceral organs, leucocytes and cultured skin fibroblasts, using either the natural substrate G_{M1}-ganglioside or the artificial substrates p-nitrophenyl-β-

D-galactopyranoside and 4-methylumbelliferyl-β-D-galactopyranoside. The degree of the deficiency varies somewhat with the substrate and the cell material used; the latter is especially important in cultured fibroblasts where the activity of β-galactosidase and some other lysosomal enzymes in control cells show considerable variations with cell cultivation conditions (Okada *et al.*, 1971; Milunsky *et al.*, 1972; Butterworth *et al.*, 1974; Heukels-Dully and Niermeijer, 1976). In patients with the infantile form of G_{M1}-gangliosidosis the β-galactosidase activity in various organs and cell types was found to vary from less than 1 to 5% of control values (O'Brien, 1972; Pinsky *et al.*, 1974; Galjaard *et al.*, 1975).

Derry and colleagues (1968) reported the first example of a juvenile form of G_{M1}-gangliosidosis in two sibs of French-Canadian ancestry. Since then other examples have been reported (O'Brien, 1972; O'Brien *et al.*, 1972). Early psychomotor development is usually normal and symptoms begin to appear between 6 and 24 months. The child starts to walk unsteadily, speech deteriorates and ataxia and muscular weakness become more evident. Subsequently, progressive mental deterioration, seizures and spasticity result in death from recurrent infections usually between ages 3 and 10 years. Unlike the infantile form there is no hepatosplenomegaly, bony deformities are mild or absent and there is no macular cherry-red spot or corneal clouding.

Classification of patients on the basis of their clinical phenotype is not always easy, as can be deduced from various reports describing patients with symptoms and signs which differ from the classical type 1 and type 2 forms of G_{M1}-gangliosidosis (Golderbert *et al.*, 1971; Feldges *et al.*, 1973; Lowden *et al.*, 1974; Pinsky *et al.*, 1974; Orii *et al.*, 1975). Histopathology does not clearly differentiate the infantile and juvenile forms, nor does the biochemical analysis of accumulated gangliosides and/or keratansulphate-like glycosaminoglycans. In neuronal tissue G_{M1}-ganglioside usually accumulates to a level where it comprises 60–80% of the total amount of ganglioside. The amount of visceral gangliosides may vary whereas the visceral keratansulphate-like mucopolysaccharides are always increased (Suzuki *et al.*, 1971). Electron microscopy of neuronal cells shows cytoplasmic inclusions of a lamellar membranous, granular or amorphous nature. Foamy vacuolization is seen in histiocytes and often in parenchymal cells of liver and spleen.

The β-galactosidase activity is decreased to a similar extent in tissues and cells from patients with either type 1 or type 2 G_{M1}-gangliosidosis. Although there have been reports that these two forms can be distinguished by different characteristics of the residual β-galactosidase activity (O'Brien *et al.*, 1972; Pinsky *et al.*, 1974), these findings seem open to doubt. Reported differences in the electrophoretic pattern, have been questioned by Suzuki and Suzuki (1974) and the kinetic differences observed by Pinsky *et al.* (1974) are, in our opinion, difficult to demonstrate with such low enzyme activities, particularly under the assay conditions described by the authors.

Yamamoto and colleagues (1974) and Loonen and co-workers (1974) reported single patients with β-galactosidase deficiency who had a relatively normal development until adulthood and then developed myoclonus, cerebellar ataxia and mental retardation. Other abnormalities including angiokeratoma corporis diffusum, macular cherry red spot and mild bone deformities were observed. The β-galactosidase activity measured in leucocytes using p-nitrophenyl substrate was 5–7% in the patient described by Yamamoto and about 10% in the patient described in our own centre (Koster et al., 1976). Residual activities of this order of magnitude were also reported in younger patients with G_{M1}-gangliosidosis (Goldberg et al., 1971; Pinsky et al., 1974).

At present, it is impossible to relate the severity of the clinical symptoms to the residual β-galactosidase since the number of patients in each category is too small and the type of material and assay conditions used vary between different authors. Even when natural substrate and well-defined material from the patient is used, it is difficult to relate the results of an enzyme assay on cell homogenates to the in vivo circumstances. Yet, it will be interesting to investigate the minimal enzyme activities required for an adequate breakdown of the cellular constituent(s) concerned (O'Brien et al., 1973). More insight into the molecular background of the aetiology and pathogenesis of the different clinical types of G_{M1}-gangliosidosis may be acquired by the elucidation of the exact structure of normal β-galactosidase and of the changes caused by different gene mutation(s).

Biochemical properties of β-galactosidase

Most studies of the biochemical properties of normal β-galactosidases have been performed on the liver enzyme. Originally, three components were separated by starch gel electrophoresis (Ho and O'Brien, 1969), Sephadex gel filtration and isoelectric focusing (Hultberg and Ockerman, 1969).

One of these forms is a neutral glycosidase which is heat stable, has a pH optimum 5.0–6.5 and a molecular weight of 45 000. There are two acidic forms with a pH optimum of 4.0–4.5; both are heat stable and stimulated by chloride ions; one of these acidic forms is macromolecular while the other has a mol wt around 70 000. Norden et al. (1974) purified the major acidic form which moves fastest in gel electrophoresis towards the anode and they found what they considered to be a single polypeptide with a mol wt of 65 000–75 000. The other more slowly migrating acidic form had a molecular weight which was about ten times greater. Immunological studies showed that antibodies against the 70 000 mol wt form cross-reacted with the macromolecular form; since the low mol wt fraction can be generated in vitro from the macromolecular form it was postulated that the latter is a multi-aggregate of the 70 000 mol wt polypeptide (O'Brien, 1975). In normal human liver studies using artificial substrate about 85% of the β-galactosidase activity at pH 4.5 is due to the low mol wt acidic form, 10% is due to the multimeric acidic form and 5% to neutral β-galactosidase.

In patients with the infantile type 1 G_{M1}-gangliosidosis and in patients with the juvenile type 2, both acidic forms of β-galactosidase are deficient whereas the activity of the neutral form varies both in patients and in controls without apparent relationship to the clinical status (O'Brien, 1975). Immunological studies indicate that inactive enzyme molecules are present in patients with type 1 and type 2 G_{M1}-gangliosidosis (O'Brien, 1975; Meisler and Ratazzi, 1974).

In cultured fibroblasts derived from the patient with the so-called type 3 G_{M1}-gangliosidosis described by Pinsky et al. (1974) and in fibroblasts from the adult patient (type 4) found by our own group (Loonen et al., 1974) we could not demonstrate any abnormalities in the kinetic properties of the residual β-galactosidase activity (Table 11.1). Dr John O'Brien (La Jolla, California) found cross reaction to anti-β-galactosidase in cultured fibroblasts from the adult type 4 patient (O'Brien et al., 1977).

Table 11.1 Characteristics of β-galactosidase in fibroblasts from patients with different clinical variants of G_{M1}-gangliosidosis

		Control	Pinsky's type 3	Adult type 4	Fusion 1 × 4
Km value (4 MU substrate)		4.2×10^{-4} m	3.5	3.9	4.4
pH Optimum (phosphate and citrate buffer)		4.4	4.4	4.4	4.4
Heat inactivation	1 h 42°C	35%	27	29	24
	1 h 50 °C	95%	84	87	89

Genetic heterogeneity studied by somatic cell hybridization

One of the questions that can be asked is whether the differences in clinical and pathological manifestations in the various types of G_{M1}-gangliosidosis are based on different gene mutations. In lower organisms complementation tests after cell fusion have significantly contributed to a better understanding of the nature of various gene mutations (Fincham, 1966). In man the first example of complementation in heterokaryons was observed after fusion of cultured fibroblasts derived from patients with different clinical forms of the autosomal recessive metabolic disease xeroderma pigmentosum (de Weerd-Kastelein et al., 1972). In cells from each of both types of patients DNA repair after ultraviolet irradiation is defective, but in heterokaryons containing the genome of both patients the DNA repair process is restored. In this case complementation analysis could be performed at the level of single binuclear heterokaryons since DNA repair can be demonstrated by autoradiography after UV irradiation and [^{3}H]thymidine incorporation.

We have tried to develop a similar procedure for the genetic analysis of β-galactosidase deficiencies in different types of G_{M1}-gangliosidosis by devising a micromethod enabling β-galactosidase assays to be made in single cultured cells (Galjaard *et al.*, 1974b, 1974c). A scheme of the procedure is illustrated in Figure 11.1.

Cultured fibroblasts from a patient with one of the types of G_{M1}-gangliosidosis are hybridized with those from another type of patient, using inactivated Sendai virus. After fusion cells were grown for 24 h in a Falcon flask, they were then trypsinized and about 4×10^4 cells were seeded on small dishes with a thin plastic bottom.

After 40 h of subsequent cultivation, the whole dish is freeze-dried overnight *in vacuo* at $-40\,^{\circ}\mathrm{C}$. At room temperature single lyophilized cells can then be isolated by free hand dissection of small pieces of the plastic bottom, using a

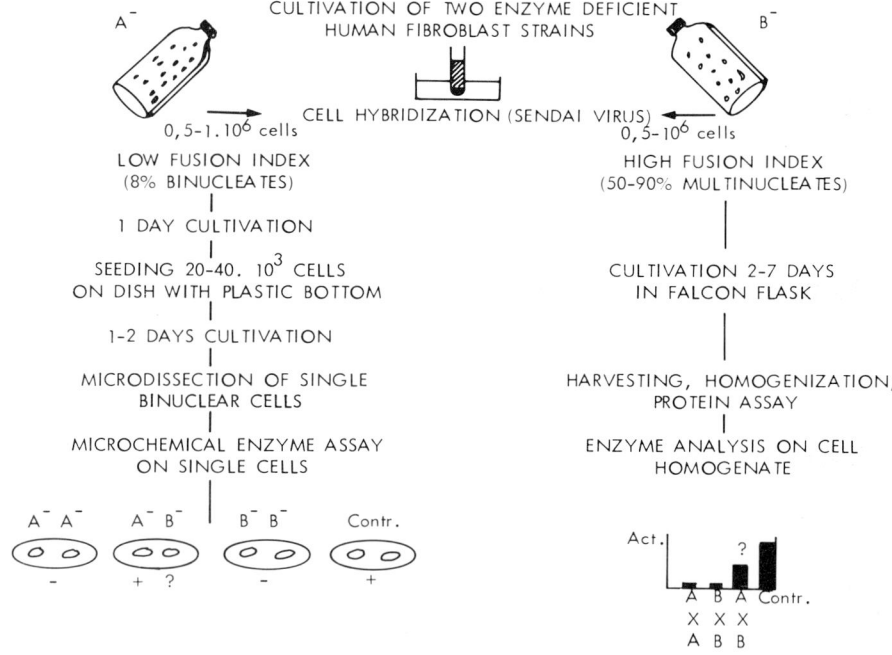

Figure 11.1   The study of genetic heterogeneity in metabolic disorders by enzyme analysis in (single) somatic cell hybrids

stereomicroscope. Each piece containing a single mononuclear fibroblast or a binuclear heterokaryon can be incubated in submicrolitre volumes of 4-methylumbelliferyl-β-D-galactopyranoside using the 'oil well technique' originally described by Lowry's group (Matschinsky *et al.*, 1968; Lowry and Passonneau, 1972). After incubation and addition of carbonate buffer pH 10.5

the fluorescence of the liberated methylumbelliferone is measured in 1 μl final volume using a microscope fluorometer. Details of the procedures for isolation of single cells and the microfluorometric measurements of their β-galactosidase activity have been described elsewhere (Galjaard et al., 1974b, 1974c).

Fusion of two human fibroblast strains under conditions which provide about 8% binuclear heterokaryons (Galjaard et al., 1974c) will yield roughly 4% binuclear cells containing the genome from each of both patients. If complementation occurs we thus expect restoration of β-galactosidase activity in 50% of the binuclear cells isolated. In each of our experiments 80–100 binuclear cells were isolated from different freeze-dried culture dishes. Activities in non-fused mononuclear cells isolated from the same culture dish serve as controls as do binuclear cells obtained from fusion of each parental cell strain with itself.

Using this procedure genetic complementation was observed after hybridization of cells from the adult type 4 G_{M1}-gangliosidosis with cells from type 1 or type 2 patients. Restoration of β-galactosidase activity also occurred when cells from Pinsky's type 3 patient were fused with type 1 or with type 2 cells. After hybridization with type 4 cells β-galactosidase activities in some of the binuclear heterokaryons were restored to control levels (Figure 11.2). Fusions between type 3 cells and type 1 or type 2 cells resulted in only a partial restoration of β-galactosidase activity (Figure 11.3). Although this phenomenon was observed in repeated experiments it should be emphasized that the interpretation depends on the β-galactosidase activity in mononuclear control fibroblasts, which is known to vary considerably between cells in the same culture and also between different cultures of the same strain (Galjaard et al., 1974d; Heukels-Dully and Niermeijer, 1976).

Complementation analysis can also be achieved by making enzyme assays on cell homogenates after hybridization under conditions such that 60–80% of the cells form multikaryons, the majority of which contains the genome of both parental cell strains (Figure 11.1). This approach is simpler and quicker than the single cell method but it enables only a qualitative rather than a quantitative answer and its sensitivity is less. Partial restoration by complementation might be missed by this less sensitive method. Yet it can be seen from Figure 11.4 that the method did allow us to differentiate between two complementation groups among the different clinical types of G_{M1}-gangliosidosis. The β-galactosidase deficiencies in Pinsky's type 3 patient and in our adult type 4 patient must be due to a different gene mutation than that involved in all other types of G_{M1}-gangliosidosis so far tested. Clinical differences between patients with the infantile and juvenile type of the disease are either caused by non-genetic factors or they are the result of different allelic mutations. The same may hold for the differences between Pinsky's type 3 and our adult type 4 patient, although definite conclusions must wait until the whole clinical course of these two patients, who are still alive, is known.

The explanation of genetic complementation after cell hybridization in cer-

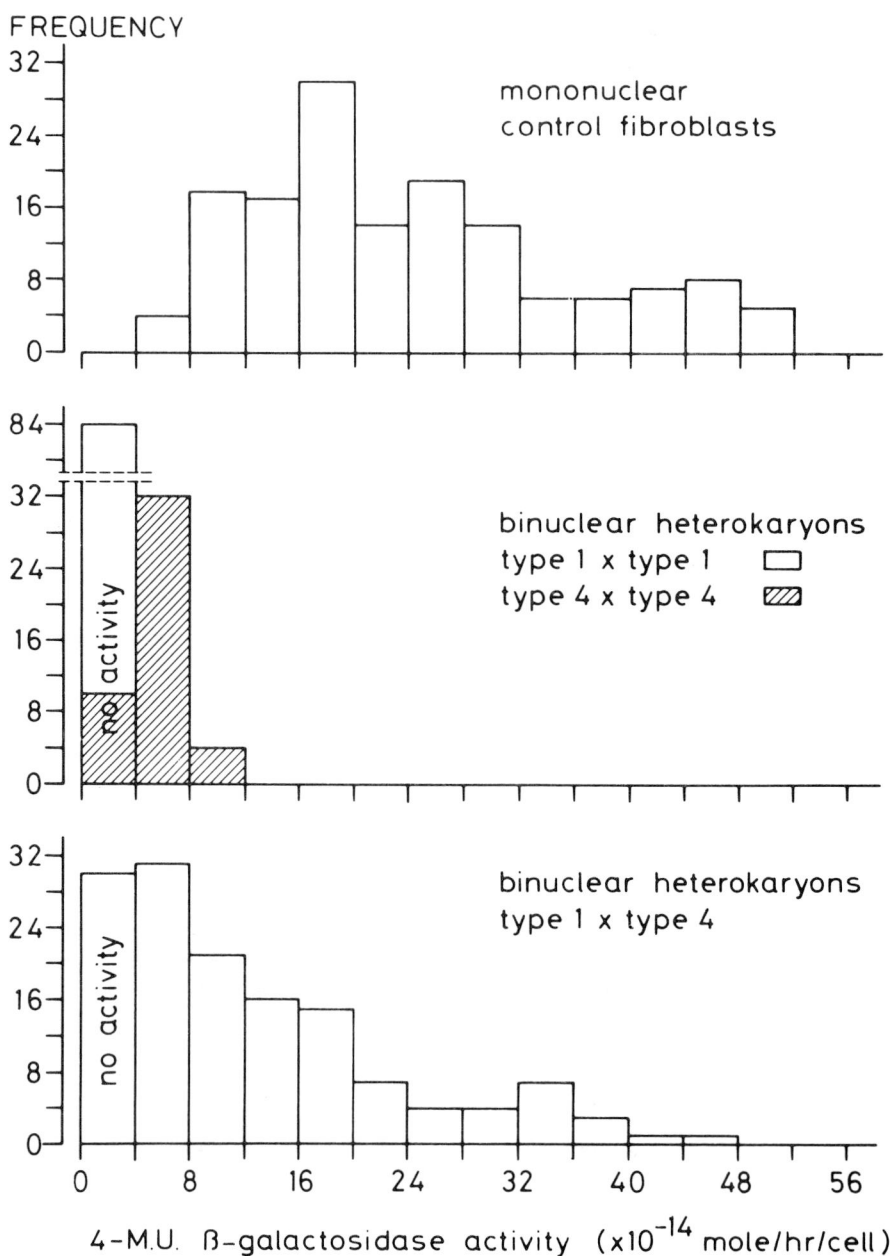

Figure 11.2 β-Galactoside in single binuclear heterokaryons after somatic cell hybridization (From Nature, *257*, 60–62 (1975) with permission)

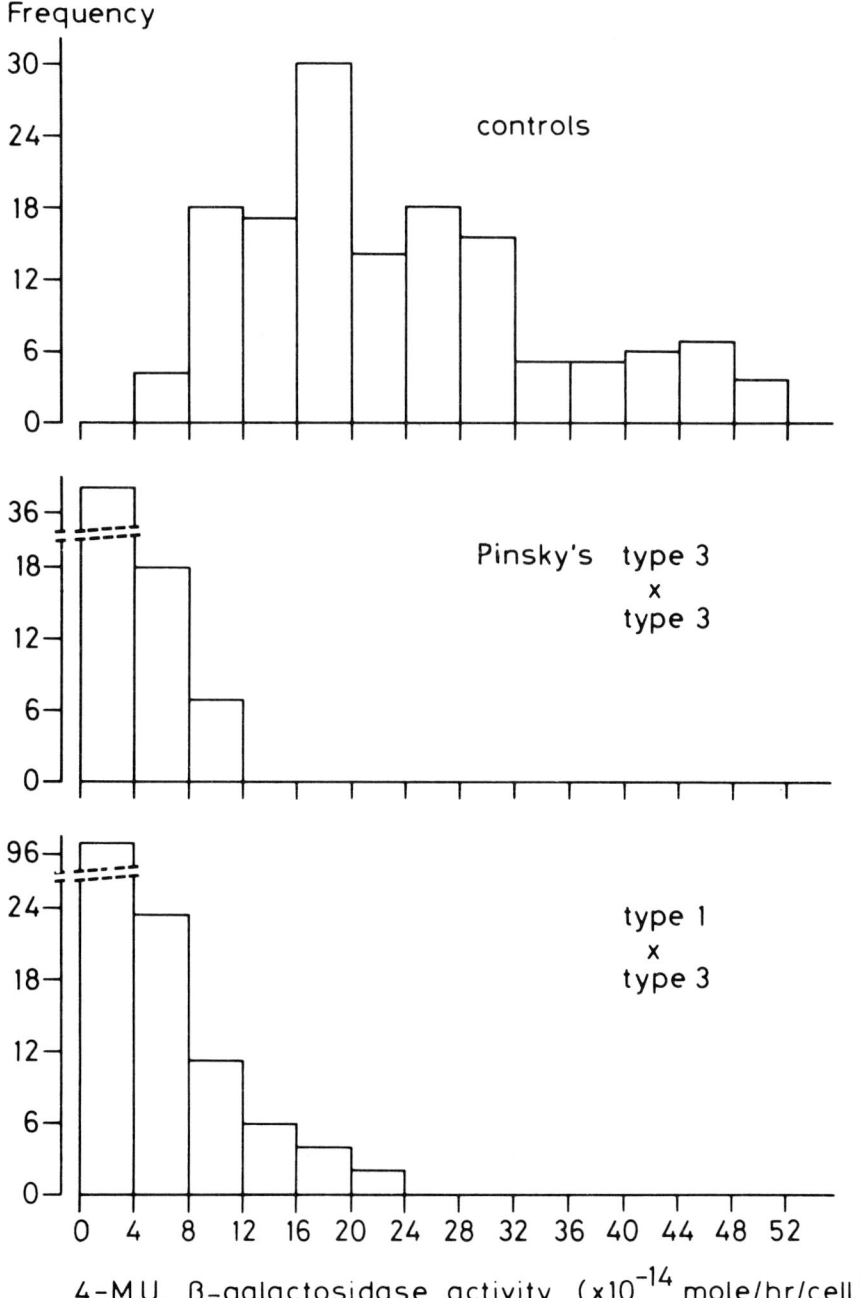

Figure 11.3 β-Galactosidase in single binuclear heterokaryons after somatic cell hybridization

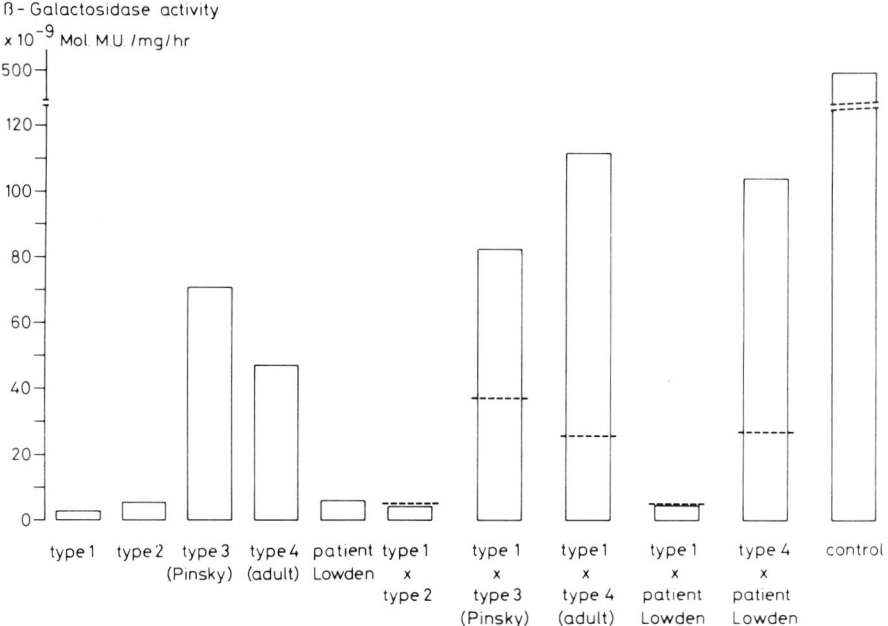

ß - Galactosidase activity

x 10^{-9} Mol. M.U. /mg/hr

Figure 11.4 Genetic complementation for β-galactosidase activity after somatic cell hybridization

tain combinations is not yet clear. According to Norden *et al.* (1974) β-galactosidase in human liver consists of a homopolymer of one single polypeptide. Intra-allelic complementation is hard to imagine although it cannot be ruled out completely. Another possibility is that in one category of patients (type 1 and type 2) a structural mutation leads to a synthesis of normal quantities of functionally inactive enzyme proteins. The residual β-galactosidase activity in type 3 and type 4 cells might be due to an insufficient production of structurally normal enzyme molecules. Our finding of normal enzyme kinetic properties of β-galactosidase in type 3 and in type 4 cells is in support of this hypothesis. A quantitative determination of the amount of cross-reacting material to anti β-galactosidase is an essential further test. In preliminary studies on fibroblasts from our adult type 4 patient O'Brien has found cross-reacting material (CRM) but the quantity of CRM is unknown. A third possible explanation of the results of our cell hybridization experiments is the existence of another protein molecule which is required to complement the 70 000 mol wt polypeptide described by Norden *et al.* (1974) and thus to give the β-galactosidase molecule its full hydrolytic activity. The enzyme deficiency in patients with type 3 and type 4 G_{M1}-gangliosidosis might be due to a structural alteration in such a protein. Experiments are underway to elucidate the molecular basis of the different β-galactosidase deficiencies and the mechanism of the genetic complementation observed.

More information about these diseases might be obtained if in addition to the molecular genetic studies described more attention were given to the pathogenesis of the different types of G_{M1}-gangliosidosis. This would require a better understanding of the way in which G_{M1}-ganglioside is degraded under normal *in vivo* conditions and for each of the β-galactosidase deficiency variants.

G_{M2}-GANGLIOSIDOSIS

Clinical and pathological manifestations

The most common form of G_{M2}-gangliosidosis has been known for a long time as Tay–Sachs disease. According to O'Brien (1972) this disease should be classified as type 1 G_{M2}-gangliosidosis, which seems less confusing than the term B variant (Sandhoff *et al.*, 1971). The first clinical signs are apparent between birth and ten months of age with apathy and delay in sitting. Hypotonia may be obvious soon after birth, whilst later there is increased muscle tone, an exaggerated extension response to sound, an inability to hold up the head and abnormal limb movements. Visual difficulties may lead to early detection of the typical macular degeneration ('cherry red spot') and peripheral blindness usually occurs after the first year of life. Seizures usually occur after the first year when psychomotor retardation becomes more and more obvious. Brain size is increased as a result of reactive proliferation of glial cells and by the age of two years the size of the cranial vault is larger than normal. Precocious puberty may develop at about this age because of hypothalamic involvement and in the terminal stages the child is quiet and hypotonic. No hepatosplenomegaly or radiological abnormalities have been observed. The disease is invariably fatal usually by the age of two to four years.

The most striking pathological manifestations in Tay–Sachs disease occur in the brain. The ganglion cells of the cerebral cortex are enormously swollen and the lysosomes are 'ballooned' because of excessive accumulation of lipid material. This results in a loss of ganglion cells and is accompanied by a proliferation of glial cells. In the cerebellum a considerable number of Purkinje cells are lost. Demyelination is consistently found and peripheral nerves as well as neurons belonging to the autonomic nervous system are affected. The typical macular changes in the retina are also due to swelling and necrosis of ganglion cells. Electron microscopy shows membranous cytoplasmic bodies up to $0.5–2.0\,\mu m$ in diameter in ganglion cells, glial cells, axon cylinders and perivascular tissue cells. These represent lysosomes filled with closely packed, concentrically arranged electrondense membranes (see Stanbury *et al.*, 1972; O'Brien, 1972; Brady and Kolodny, 1972).

Studies by the groups of Svennerholm, Gatt, Brady and Jatzkevitz and others, have led to the identification of the storage product and to the elucidation of the genetic metabolic defect. Svennerholm (1966) was the first to suggest that G_{M2}-ganglioside storage in Tay–Sachs disease was related to a hexosaminidase deficiency. Robinson and Stirling (1968) then identified two isoenzymes with an acidic pH optimum and Okada and O'Brien (1969) finally demonstrated that Tay–Sachs disease was caused by a deficiency of one of these isoenzymes, β-D-N-acetylhexosaminidase A. As is illustrated in Figure 11.5 this isoenzyme is thought to hydrolyse G_{M2}-ganglioside and its asialo derivative (Sandhoff et al., 1971). A genetic deficiency of hexosaminidase A (hex A) results in accumulation of G_{M2}-ganglioside and of asialoganglioside in brain and to a lesser extent in visceral organs.

Tay–Sachs disease has an autosomal recessive inheritance; in a population of Ashkenazi Jews the frequency of heterozygous carriers is relatively high (1:30) whereas in non-Jewish populations or in Sephardic Jews the heterozygote frequency is only 1:300. Ashkenazi Jewish couples have therefore a 100 times higher risk of having affected offspring than other couples. This was the reason for large scale screening for heterozygous carriers in Ashkenazi Jews using automated procedures for the detection of decreased levels of hex A activity in serum (see Kaback et al., 1974). Since hex A deficiency can also be demonstrated in cultured amniotic fluid cells couples at risk can ask for prenatal diagnosis. If an affected fetus is detected the pregnancy can be interrupted and thus prevent the birth of a seriously handicapped child. When normal enzyme values are found, parents can be reassured and continue their pregnancy (O'Brien, 1973, 1975).

In 1968 Sandhoff, Andreae and Jatzkewitz described an exceptional case of G_{M2}-gangliosidosis where both acidic hexosaminidases were absent. Later, several of these patients were described (Sandhoff et al., 1971; O'Brien et al., 1972). According to O'Brien these patients should be classified as type 2 G_{M2}-gangliosidosis but they are also designated as type 0 variant or Sandhoff disease. Patients present the clinical picture of Tay–Sachs disease with onset of symptoms by 9 months of life, progressive blindness, spastic paraplegia, macrocephaly and decerebrate rigidity. Most patients die between 2 and 3 years of age. Pathological manifestations are neuronal lipidosis throughout the central and autonomic nervous systems. Histiocytes with vacuolated cytoplasm due to lysosomes filled with storage products are found in visceral organs and vacuolization is also present in neural tubular epithelium. Accumulation of G_{M2}-ganglioside and asialogangliosides in the affected brain may produce concentrations up to 300 times and 100 times respectively the values found in control brains. Storage of these products also occurs in visceral organs although to a lesser degree. Unlike Tay–Sachs disease there are accumulations of globosides in liver, spleen and kidney. This must be due to the absence of hexosaminidase B (hex B) activity which isoenzyme is normally involved in the hydrolysis of terminal β-glycosidically bound N-acetylgalactosamine in asialo-G_{M2}-ganglioside and in globoside (Figure 11.5). These two components can also be degraded by hex A.

Figure 11.5 Schematic representation of ganglioside degradation

A juvenile type of G_{M2}-gangliosidosis has been reported by several authors (O'Brien, 1972) and it has been classified as type 3. These patients develop progressive psychomotor retardation and ataxia between the 2nd and 5th year of life and death occurs between 5 and 15 years. Macrocephaly and macular degeneration have not been observed but optic atrophy may be present. G_{M2}-ganglioside accumulation is less pronounced and this might be related to the residual activity of hex A, which has been demonstrated by Okada *et al.* (1970).

Finally a number of exceptional cases have been described in whom it is difficult to relate the clinical phenotype to the enzymological data. Sandhoff and co-workers (1971) described an individual whom they called 'AB variant'; no clinical signs were present, hex A and B activities were higher than normal and G_{M2}-ganglioside and its asialo derivative nevertheless accumulated in brain. Spence and colleagues (1974) described a patient with type 2 G_{M2}-gangliosidosis in whom hex A and hex B activities were reduced to 7–10% in leucocytes and cultured fibroblasts and to 20–24% in plasma. This patient showed psychomotor retardation 1 year after birth and had seizures and retinal degeneration at 2 years. Such findings make it difficult to explain observations of healthy individuals where hex A activity is absent (Navon *et al.*, 1973) or where hex B is deficient (Dreyfus *et al.*, 1975). In both instances these individuals were related to patients with G_{M2}-gangliosidosis and it has been suggested that they might be heterozygous for two mutations, both involved in hexosaminidases. It should, however, be emphasized that hexosaminidase activities in both instances were measured with artificial methylumbelliferyl- or *p*-nitrophenyl substrates. The activities measured in this way and under *in vitro* conditions do not necessarily reflect the ability of living nerve cells to degrade gangliosides. Thus far the assay for hexosaminidase A using the natural substrates, G_{M2}-ganglioside or its asialo derivative has not been very successful. In the vast majority of cases, however, the use of artificial substrates has been of great help in the diagnosis of patients with Tay–Sachs or Sandhoff disease, for the screening of heterozygous carriers and for prenatal diagnosis in early pregnancy.

Biochemical properties of hexosaminidases

Electrophoresis (Robinson and Stirling, 1968; Okada and O'Brien, 1969), isoelectric focusing (Hultberg, 1969; Sandhoff et al., 1971) and DEAE cellulose column chromatography (Ellis et al., 1975) have been used to investigate the isoenzymes of hexosaminidase. Originally two isoenzymes hex A and hex B were identified, both with a pH optimum of 4.4 and a K_m of 5×10^{-4} M. Hex A has a faster mobility towards the anode, is heat labile (5–10 min at 50 °C destroys its activity) and has a molecular weight of about 100 000. Hex B moves more slowly to the anode in cellulose acetate electrophoresis, is heat stable and has a similar molecular weight. It is thought that hex A alone cleaves the terminal N-acetylgalactosamine from G_{M2}-ganglioside whereas both hex A and hex B are able to hydrolyse the β-glycosidically bound N-acetyl-galactosamine from asialo G_{M2}-ganglioside and globoside (Figure 11.5).

Hex A in blood plasma has a faster mobility on electrophoresis than tissue hex A. The former can be converted into hex B by neuraminidase and by other non-specific treatments in tissue (Swallow et al., 1974). Beutler and Kuhle (1975) were able to generate hex B from hex A by merthiolate treatment or by repeated freezing and thawing. Immunological studies also suggest a common subunit in hex A and hex B because antibodies against purified hex B cross-react with hex A (Bartholomew and Rattazzi, 1974; Beutler et al., 1975).

On the basis of chemical and immunological studies several models have been proposed for the molecular composition of normal hex A and hex B. Gene localization studies using rodent–human somatic cell hybrids have increased the confusion since various investigators came to different conclusions about the interdependency of the gene loci for hex A and hex B (van Someren et al., 1973; Lalley et al., 1974; Gilbert et al., 1975). This confusion appears to have been caused by a misinterpretation of certain bands in electrophoresis, which our group has now identified as hex S and as heteropolymers between human specific hex A subunits and rodent hexosaminidase subunits (van Cong et al., 1975; Hoeksema et al., 1977; Reuser and Galjaard, 1976).

At present it seems to be established that in interspecies somatic cell hybrids hex A cannot be expressed in the absence of hex B. Hex B activity on the other hand can be demonstrated in hybrids which have lost the gene locus specific for hex A. The gene locus specific for hex A has now been assigned to chromosome 15 (van Heijningen et al., 1974; Lalley et al., 1975) and the locus for hex B to chromosome 5 (Gilbert et al., 1975).

Another source of confusion in the investigation of hexosaminidase isoenzymes has been the presence of both hex C and hex S. The former was first discovered in brain tissue by Hooghwinkel *et al.* (1972). More detailed studies have shown that hex C has a faster mobility in gel electrophoresis than hex A, its pH optimum is about 7.0 and unlike hex A and hex B is localized in the cytoplasm rather than in the lysosomes. Antisera against placental hex A or against a mixture of hex A and hex B do not cross-react with this neutral hex C (Braidman *et al.*, 1974; Penton *et al.*, 1975). No structural relationships have yet been established between the neutral hexosaminidase C and the acidic forms hex A and hex B. The reason why hex C was not detected earlier is its lability, even at low temperature.

In Tay–Sachs disease hex A is deficient (Okada and O'Brien, 1969) and hex B activity is usually increased. The heat stability, pH optimum and Michaelis constants for hex B in patients with Tay–Sachs disease and in normal individuals are similar (Sandhoff *et al.*, 1971). Some investigators do not find any residual activity of hex A in tissues or cultured cells from Tay–Sachs patients, but others report residual activities of a few per cent. The use of heat inactivation in the assay procedure for hex A activity presents difficulties since some inactivation of other hexosaminidase components might easily be misinterpreted as representing hex A activity.

Using a method which enables a clear distinction between hex S and hex C in cells from all patients with Tay–Sachs disease so far tested we have found hex C activity (Reuser and Galjaard, 1976).

In cells and tissues of patients with Sandhoff disease there is a marked deficiency of both hex A and hex B although a small residual activity for both acidic isoenzymes can usually be demonstrated. In some electrophoretic studies of cell material from Sandhoff patients a clear band is seen moving faster toward the anode than hex A (Braidman *et al.*, 1974; Galjaard *et al.*, 1974a). This band has been designated as hex C but it also has definite activity at pH 4.4. Ikonne and colleagues (1975) and Beutler and co-workers (1975) have characterized this fast anodal moving band as hex S. It involves most of the residual hexosaminidase activity in Sandhoff patients. Its pH optimum is about 4 and it cross reacts with antisera prepared against hex A but not with anti hex B (Beutler *et al.*, 1975a). Using a method enabling a clear distinction to be made between the acidic hex S and the neutral hex C, we could not demonstrate hex S in cells from Tay–Sachs patients whereas it was clearly present in control cells (Reuser and Galjaard, 1976).

In a series of elegant experiments, Beutler and Kuhl (1975) generated *in vitro* both hex B and hex S from hex A purified from human placenta. By separating the subunits in a mixture of hex B and hex S they were also able to generate hex A which apparently consists of different subunits that occur in hex B and hex S. The model, proposed earlier by Ropers and Schwantes (1973), has now been proven and extended by these chemical, immunological and somatic cell genetic studies. Hex A is thought to be a polymer of two different polypeptides $\alpha\beta$; hex B is a homopolymer $\beta\beta$ and hex S is a homopolymer of α-units. The neutral form, hex C, probably has no structural relationships with these three acidic forms of hexosaminidase. The gene locus for the α subunit is localized on chromosome 15 and the gene coding for the β subunit is assigned to chromosome 5.

The nature of gene mutations in variants of G_{M2}-gangliosidosis

In order to investigate the genetic background of the variants of G_{M2}-gangliosidosis we have performed complementation analysis after somatic cell hybridization of cultured skin fibroblasts derived from patients with different types of G_{M2}-gangliosidosis. We have followed the scheme illustrated in Figure 11.1. Cells (10^6) from a patient with Tay–Sachs disease were fused with 10^6 cells from a patient with Sandhoff disease using conditions yielding 60–80% multikaryons. About 40 h after fusion, cells were harvested and the hex A activity was analysed in the cell homogenate after determination of its protein content. The results were compared with those of fusions of each of the parental cell strains with itself. Equal numbers of Tay–Sachs and Sandhoff cells were also co-cultivated for 40 h without fusion and the hex A activity was then measured in the mixed cell population.

Hex A activity was determined by comparing the hydrolytic activity before and after heat inactivation using 4-methylumbelliferyl-2-acetamido-2-deoxy-β-D glucopyranoside (4-MU) as a substrate at pH 4.4. In addition electrophoresis on cellulose acetate was performed at pH 6.0 for 45 min at 20 °C. After 30 min incubation with 4-MU substrate at pH 4.5 followed by spraying with carbonate buffer (pH 10.5) the isoenzymes pattern was photographed.

The fraction of heat labile hexosaminidase activity at low pH (mainly hex A) is about 50% of the total activity in control fibroblasts. The process of cell hybridization does not affect this percentage of hex A activity. In fusions of Tay–Sachs cells with themselves, only a small fraction (about 14%) is heat labile. The total hexosaminidase activity in Sandhoff cells was too low for a reliable measurement of separate hex A and hex B activities. In the mixed population of Tay–Sachs and Sandhoff cells after two days of co-cultivation 14% hex A activity was found and this was produced by the Tay–Sachs cells whose total hex activity is even higher than that of control fibroblasts. However, after cell hybridization of Tay–Sachs with Sandhoff cells the percentage of heat labile hexosaminidase clearly increased (up to 27%). Details of these experiments have been described elsewhere (Galjaard *et al.*, 1974a).

Electrophoretic characterization (Figure 11.6) showed that somatic cell hybridization between Tay–Sachs and Sandhoff cells resulted in the appearance of hex A activity which is absent in both parental cell strains. The absence of hex A activity in the mixed cell population after two days co-cultivation of both cell strains indicates that genetic interaction is required for the restoration of hex A activity. Similar results were obtained independently

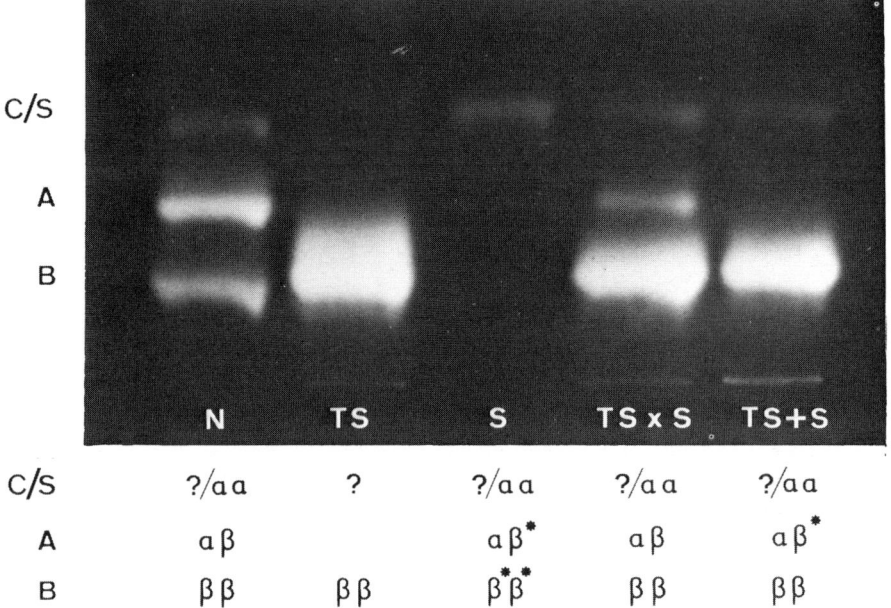

Figure 11.6 Hexosaminidase isoenzyme pattern: electrophoresis and subunit model. Separation on cellulose acetate gels. Running buffer: 50 mM potassium phosphate pH 6.8. A, B, C/S represent the different isoenzymes. N: control human fibroblasts. TS: fibroblasts from a Tay–Sachs patient. S: fibroblasts from a Sandhoff patient. TS × S: heterokaryons from Tay–Sachs × Sandhoff fibroblasts. TS + S: mixture of Tay–Sachs and Sandhoff fibroblasts after co-cultivation

by Thomas and colleagues (1974) and later by several other groups (Ropers *et al.*, 1975).

The results of these cell genetic studies prove that the enzyme deficiencies in Tay–Sachs and Sandhoff disease are based on two different gene mutations. The most likely explanation is that Tay–Sachs disease is caused by a gene mutation in the polypeptide specific for hex A (α chain) whereas Sandhoff disease is based on a mutation in the common β subunit of hex A and hex B (β chain). The residual hex A and hex B activity present in Sandhoff cells indicate that defective β chains are produced and that these are capable of association with normal α chains. Probably this binding is less than normal since the association of α chains among themselves might explain the relatively high activity of hex S in Sandhoff cells.

Cell fusion between human Sandhoff cells and cells from Chinese hamster and the pattern of heteropolymer formation indicates that in Sandhoff cells normal α chains are available (van Cong *et al.*, 1975; Hoeksema *et al.*, 1977). Immunological studies on different tissues and cells from patients with Sandhoff disease have not yielded conclusive results about the presence of material which cross-reacts with anti hex A and/or hex B antiserum.

In Tay–Sachs disease the defective α chain does not seem capable of binding with the normal β chains, since no residual hex A activity is found on electrophoresis even when sensitive detection methods are used. Again, immunological studies are not completely clear because some workers do not find any cross reaction with anti hex A antibodies (Bartholomew and Rattazzi, 1974) but Carrol and Robinson (1973) did find cross-reacting material in liver tissue. The absence of association between defective α chains and normal β chains leaves the latter free for the formation of more hex B ($\beta\beta$ chains) and thus probably accounts for the consistently higher hex B activity in Tay–Sachs cells.

The mechanism of genetic complementation after fusion of Tay–Sachs with Sandhoff cells resulting in restoration of hex A activity can be explained by the association of normal α chains produced by Sandhoff cells with normal β chains synthesized by Tay–Sachs cells (Figure 11.6). Complementation analyses at different time intervals after cell fusion and after inhibition of protein synthesis suggest that restoration of hex A activity in heterokaryons is mainly the result of *de novo* protein synthesis. The hex A formed as a result of genetic complementation has all the properties of normal isoenzyme. It is for instance taken up by deficient Sandhoff cells in co-cultivation experiments as is the case when normal fibroblasts are co-cultivated with Sandhoff cells (Reuser *et al.*, 1976).

Thus far we have hybridized four different Sandhoff cell strains with five different cell lines from patients with Tay–Sachs disease, some of which were clinical variants. In all Tay–Sachs × Sandhoff fusions genetic complementation occurred. In no instance was restoration of hex A activity observed when cells from different patients with Tay–Sachs disease were hybridized; the same was true for different Sandhoff cell strains. The clinical and/or biochemical differences within each category of G_{M2}-gangliosidosis may thus be due to different mutations affecting the same subunit of the enzyme. Although the molecular structure of the hexosaminidase isoenzymes has been elucidated and more insight has been gained into the nature of the gene mutations, there is still much to be learned about the exact *in vivo* function of the various isoenzymes. Only a better understanding of these functions in the various normal tissues will help us to explain the wide variations in clinical phenotype with the same enzyme deficiency and the discrepancies between clinical manifestations and enzyme activities described in a number of cases (Navon *et al.*, 1973; Dreyfus *et al.*, 1975).

References

Bartholomew, W. R. and Rattazzi, M. C. (1974). Immunochemical characterization of human hexosaminidase from normal individuals and patients with Tay–Sachs disease. *Int. Arch. Allergy*, **46**, 512

Beutler, E., Kuhl, W. and Comings, D. (1975). Hexosaminidase isozyme in type 0 G_{M2}-gangliosidosis. *Am. J. Hum. Genet.*, **27,** 628

Beutler, E. and Kuhl, W. (1975). Subunit structure of human hexosaminidase verified: interconvertibility of hexosaminidase iso-enzymes. *Nature (London)*, **258,** 263

Brady, R. O. and Kolodny, E. H. (1972). Disorders of ganglioside metabolism. *In*: A. G. Steinberg and A. G. Bearn (eds.). *Progress in Medical Genetics,* Vol. 8, pp. 225–241. (New York: Grüne and Stratton)

Braidman, J., Carroll, M., Dance, N., Robinson, D., Poenaru, L., Weber, A., Dreufus, J. C., Overdijk, B. and Hooghwinkel, G. J. M. (1974). Characterization of human hexosaminidase C. *FEBS Lett.,* **41,** 181

Butterworth, J., Sutherland, G. R., Broadhead, D. M. and Bain, A. D. (1974). Effect of serum concentration, type of culture medium and pH on the lysosomal enzyme activity of cultured human amniotic fluid cells. *Clin. Chim. Acta,* **53,** 239

Carroll, M. and Robinson, D. (1973). Immunological properties of N-acetyl-β-D-glucosaminidase of normal human liver and of G_{M2}-gangliosidosis liver. *Biochem. J.,* **131,** 91

Cong, N. V., Weil, D., Rebourcet, R. and Frézal, J. (1975). A study of hexosaminidases in interspecific hybrids and in G_{M2}-gangliosidosis. *Ann. Hum. Genet.,* **39,** 111

Derry, D. M., Fawcett, J. S., Andermann, F. and Wolfe, L. S. (1968). Late infantile systemic lipidosis. *Neurology,* **18,** 340

Dreyfus, J. C., Poenaru, L. and Svennerholm (1975). Absence of hexosaminidase A and B in a normal adult. *N. Engl. J. Med.,* **292,** 61

Ellis, R. B., Ikonne, J. U. and Mason, P. K. (1975). DEAE cellulose microcolumn chromotography coupled with automated assay for hexosaminidase components. *Anal. Biochem.,* **63,** 5

Feldges, A., Müller, H. J., Bühler, E. and Stalder, G. (1973). G_{M1}-gangliosidosis, clinical aspects and biochemistry. *Helv. Paediatr. Acta,* **28,** 511

Fincham, J. R. S. (1966). Genetic Complementation. (New York: W. A. Benjamin Inc.)

Galjaard, H., Hoogeveen, A., de Wit-Verbeek, H. A., Reuser, A. J. J., Keijzer, W., Westerveld, A. and Bootsma, D. (1974a). Tay–Sachs and Sandhoff's disease, intergenic complementation after somatic cell hybridization. *Exp. Cell Res.,* **87,** 444

Galjaard, H., Hoogstraten, J. J. van, Josselin de Jong, J. E. and Mulder, M. P. (1974b). Methodology of the quantitative cytochemical analysis of single cultured cells. *Histochem. J.,* **6,** 409

Galjaard, H., Hoogeveen, A., Keijzer, W., de Wit-Verbeek, H. A. and Vlek-Noot (1974c). The use of quantitative cytochemical analyses in rapid prenatal detection and somatic cell genetic studies of metabolic diseases. *Histochem. J.,* **6,** 491

Galjaard, H., Reuser, A. J. J., Heukels-Dully, M., Hoogeveen, A., Keijzer, W., de Wit-Verbeek, H. A. and Niermeijer, M. F. (1974d). Genetic heterogeneity and variation of lysosomal enzyme activities in cultured human cells. *In*: J. M. Tager, G. J. M. Hooghwinkel and W. Th. Daems (eds.) pp. 35–51. *Enzyme Therapy in Lysosomal Storage Diseases.* (Amsterdam: North Holland Publishing Co.)

Galjaard, H., Hoogeveen, A., Keijzer, W., de Wit-Verbeek, H. A., Reuser, A. J. J., Mae Wan Ho and Robinson, D. (1975). Genetic heterogeneity in G_{M1}-gangliosidosis. *Nature (London)*, **257,** 60

Gilbert, F., Kucherlapati, R. P., Creagan, R. P., Murnane, M. J., Darlington, G. J. and Ruddle, F. H. (1975). The assignment of genes for hexosaminidase A and B to individual chromosomes. *Proc. Nat. Acad. Sci. U.S.A.,* **72,** 263

Goldberg, M. F., Cotlier, E., Fichenscher, L. G., Kenyon K. (1971). Macular cherry-red spot, corneal clouding and β-galactosidase deficiency. *Arch. Int. Med.,* **128,** 387

Gravel, R. A., Mahoney, M. J., Ruddle, F. H. and Rosenberg, L. E. (1975). Genetic complementation in heterokaryons of human fibroblasts defective in cobalamin metabolism. *Proc. Nat. Acad. Sci. U.S.A.,* **72,** 3181

Harris, H. (1976). *The Principles of Human Biochemic Genetics*. 3rd ed. (Amsterdam: North Holland Publishing Co.)

Heukels-Dully, M. J. and Niermeijer, M. F. (1976). Variation in lysosomal enzyme activity during growth in culture of human fibroblasts and amniotic fluid cells, *Exp. Cell Res.*, **97**, 304

Heijningen, V. van, Bobrow, M., Bodmer, W. F., Gardiner, S. E., Povey, S. and Hopkinson, D. A. (1975). Chromosome assignment of some human enzyme loci. *Ann. Hum. Genet.*, **38**, 295

Ho, M. W. and O'Brien, J. S. (1969). Hurler's syndrome: deficiency of a specific beta galactosidase isoenzyme. *Science*, **165**, 611

Hoeksema, H. L., Reuser, A. J. J., Hoogeveen, A., Westerveld, A., Braidman, J. and Robinson, D. (1977). Characterization of hexosaminidase isoenzymes in man–Chinese hamster somatic cell hybrids. *Am. J. Hum. Genet.* **29**, 14–23

Hooghwinkel, G. J. M., Veltkamp, W. A., Overdijk, B. and Lisman, J. J. W. (1972). Electrophoretic separation of hexosaminidases of human and bovine brain and liver. *Hoppe-Seyler's Z. Physiol. Chem.*, **353**, 839

Hultberg, B. (1969). *N*-acetyl hexosaminidase activities in Tay–Sachs disease. *Lancet*, **ii**, 1195

Hultberg, B. and Ockerman, P. A. (1969). Properties of 4-methylumbelliferyl β-galactosidase activities in human liver. *Scand. J. Clin. Lab. Invest.*, **23**, 213

Human Gene Mapping: In: Rotterdam Conference (1974), 2nd International Workshop on Human Gene Mapping. *Birth Defects: Original Article Series*, Vol. XI, no. 3 (The National Foundation, New York, 1975)

Human Gene Mapping: In: Baltimore Conference (1975). 3rd International Workshop on Human Gene Mapping. *Birth Defects: Original Article Series*, Vol. XII, no. 7 (The National Foundation, New York, 1976)

Ikonne, J. U., Rattazzi, M. C. and Desnick, R. J. (1975). Characterization of hex S, the major residual β-hexosaminidase activity in type 0 G_{M2}-gangliosidosis. *Am. J. Hum. Genet.*, **27**, 639

Kaback, M. R. S., Zeiger, R. S., Reynolds, L. W. and Sonneborn, H. (1974). Approaches to the control and prevention of Tay–Sachs disease. *In*: A. G. Steinberg and A. G. Bearn (eds.). *Progress in Medical Genetics*, Vol. 10, pp. 103–134. (New York: Grüne and Stratton Inc.)

Kirkman, H. N. (1972). Enzyme Defects. *In*: A. G. Steinberg and A. G. Bearn (eds.). *Progress in Medical Genetics*, **8**, pp. 125–167. (New York: Grüne and Stratton, Inc.)

Koster, J. F., Niermeijer, M. F., Loonen, M. C. B. and Galjaard, H. (1976). β-galactosidase deficiency in an adult: a biochemical and somatic cell genetic study on a variant of G_{M1}-gangliosidosis, *Clin. Genet.*, **9**, 427

Kraemer, K. H., De Weerd-Kastelein, E. A., Robbins, J. H., Keijzer, W., Barrett, S. F., Petinga, R. A. and Bootsma, D. (1975). Five complementation groups in xeroderma pigmentosum, *Mutat. Res.*, **33**, 327

Lalley, P. A., Rattazzi, M. C. and Shows, T. D. (1974). Human hexosaminidase A and B, expression and linkage relationships in somatic cell hybrids. *Proc. Nat. Acad. Sci. U.S.A.*, **71**, 1569

Loonen, M. C. B., Lugt, L. van de and Franke, C. L. (1974). Angiokeratomata corporis diffusion and lysosomal enzyme deficiency. *Lancet*, **ii**, 785

Lowden, J. A., Callahan, J. W., Norman, M. G., Thain, M. and Prichard, J. S. (1974). Juvenile G_{M1}-gangliosidosis: occurrence with absence of two β-galactosidase components. *Arch. Neurol.*, **31**, 200

Lowry, O. H. and Passonneau, J. V. (1972). *A Flexible System of Enzymatic Analysis*, pp. 129–145. (New York and London: Academic Press)

Lyons, L. B., Cox, R. P. and Dancis, J. (1973). Complementation analysis of maple syrup urine disease in heterokaryons derived from cultured human fibroblasts. *Nature (London)*, **243**, 533

Matschinsky, F. M., Passonneau, J. V. and Lowry, O. H. (1968). Quantitative histochemical analysis of glycolytic intermediates and cofactors with an oil well technique. *J. Histochem. Cytochem.*, **16**, 29

McKusick, V. A., Howell, R. R., Hussels, J., Neufeld, E. F. and Stevenson, R. E. (1972). Allelism, non-allelism and genetic compounds among the mucopolysaccharidoses. *Lancet*, **ii**, 993

McKusick, V. A. (1975). *Mendelian Inheritance in Man*. 4th ed. (Baltimore: Johns Hopkins University Press)

Meisler, M. and Rattazzi, M. C. (1974). Immunological studies of β-galactosidase in normal human liver and G_{M1}-gangliosidosis. *Am. J. Hum. Genet.*, **26**, 683

Milunsky, A., Spielvogel, C. and Kanfer, J. N. (1972). Lysosomal enzyme variations in cultured normal skin fibroblasts. *Life Sci.*, **11**, 1101

Nadler, H. (1972). Tissue culture and antenatal detection of molecular diseases. *Biochemistry*, **54**, 677

Navon, R., Padeh, B. and Adam, A. (1973). Apparent deficiency of hexosaminidase A in healthy members of a family with Tay–Sachs disease. *Am. J. Hum. Genet.*, **25**, 287

Neufeld, E. F. (1974). The biochemical basis for mucopolysaccharidoses and mucolipidoses. *In*: A. G. Steinberg and A. G. Bearn (eds.). *Progress in Medical Genetics*, Vol. 10, pp. 81–102. (New York: Grüne and Stratton)

Nitowsky, H. M. (1972). Use of cell culture techniques in diagnosis and studies of inherited diseases. *Semin. Hematol.*, **9**, 403

Norden, A. G. W., Tennant, L. L. and O'Brien, J. S. (1974). G_{M1}-ganglioside β-galactosidase A. *J. Biol. Chem.*, **249**, 7969

O'Brien, J. S. (1972). *In*: H. Harris and K. Hirschhorn (eds.). *Advances in Human Genetics*, Vol. 3, pp. 39–98. (New York: Plenum Press)

O'Brien, J. S., Ho, M. W., Veath, M. L. *et al.* (1972). Juvenile G_{M1}-gangliosidosis, clinical, pathological, chemical and enzymatic studies. *Clin. Genet.*, **3**, 411

O'Brien, J. S. (1973). Tay–Sachs disease: from enzyme to prevention. *Fed. Proc. Fed. Am. Soc. Exp. Biol.*, **32**, 191

O'Brien, J. S., Miller, A. L., Loverde, A. W. and Veath, M. L. (1973). Sanfilippo disease type B: enzyme replacement and metabolic correction in cultured fibroblasts. *Science*, **181**, 753

O'Brien, J. S. (1975). Molecular genetics of G_{M1}-β-galactosidase. *Clin. Genet.*, **8**, 303

O'Brien, J. S. and Norden, H. G. W. (1977). Nature of the mutation in adult β-galactosidase deficient patients. *Am. J. Hum. Genet.*, **29**, 184

Okada, S. and O'Brien, J. S. (1968). Generalized gangliosidosis, β-galactosidase deficiency. *Science*, **160**, 1002

Okada, S., Veath, L. M., Leroy, J. and O'Brien, J. S. (1971). Ganglioside G_{M2} storage diseases: hexosaminidase deficiencies in cultured fibroblasts. *Am. J. Hum. Genet.*, **23**, 55

Okada, S. and O'Brien, J. S. (1969). Tay–Sachs disease. Generalized absence of a β-D-N-acetylhexosaminidase component. *Science*, **165**, 698

Okada, S., Veath, L. M. and O'Brien, J. S. (1970). Juvenile G_{M2}-gangliosidosis, partial deficiency of hexosaminidase A. *J. Pediatr.*, **77**, 1063

Orii, T., Sekegawa, K. and Kudoh, T. (1975). A variant of G_{M2}-gangliosidosis type 2. *Tohoku J. Exp. Med.*, **117**, 99

Penton, E., Poenaru, L. and Dreyfus, J. S. (1975). Hexosaminidase C in Tay–Sachs and Sandhoff disease. *Biochim. Biophys. Acta*, **391**, 162

Pinsky, L., Miller, J., Standfield, B., Watters, G. and Wolfe, L. (1974). G_{M1}-gangliosidosis in skin fibroblast culture: enzymatic differences between types 1 and 2 and observations on a third variant. *Am. J. Hum. Genet.*, **26**, 563

Reuser, A. J. J. and Galjaard, H. (1976). Characterization of hexosaminidase C and S in fibroblasts from control individuals and patients, with Tay–Sachs disease. *FEBS Lett.*, **71**, 1

Reuser, A. J. J., Halley, D., de Wit-Verbeek, H. A., Hoogeveen, A., van der Kamp, M., Mulder, M. P. and Galjaard, H. (1976). Intercellular exchange of lysosomal enzymes: enzyme assays in single human fibroblasts after co-cultivation. *Biochem. Biphys. Res. Commun.*, **69**, 311

Robinson, D. and Stirling, J. L. (1968). N-acetyl-β-D-glucosaminidases in human spleen. *Biochem. J.*, **107**, 321

Ropers, H. H. and Schwantes, U. (1973). On the molecular basis of Sandhoff's disease. *Humangenetik*, **20**, 167

Ropers, H. H., Grzeschik, K. H. and Bühler, E. (1975). Complementation after fusion of Sandhoff and Tay–Sachs fibroblasts. *Humangenetik,* **26,** 117

Sandhoff, K., Andreae, U. and Jatzkewitz, H. (1968). Deficient hexosaminidase activity in an exceptional case of Tay–Sachs disease with additional storage of kidney globoside in visceral organs. *Life Sci.,* **7,** 283

Sandhoff, K., Harzer, K., Wässle, W. and Jatzkewitz H. (1971). Enzyme alterations and lipid storage in three variants of Tay–Sachs disease. *J. Neurochem.,* **18,** 2469

Spence, M. W., Ripley, B. A., Embil, J. A. and Tribbles, J. A. R. (1974). A new variant of Sandhoff's disease. *Pediatr. Res.,* **8,** 628

Stanbury, J. B., Wijngaarden, J. B. and Frederickson, D. S. (1972). *The Metabolic Basis of Inherited Disease.* 3rd ed. (New York: McGraw Hill Book Co.)

Suzuki, Y. and Suzuki, K. (1974). Glycosphingolipid β-galactosidases IV. *J. Biol. Chem.,* **249,** 2113

Suzuki, Y., Crocker, A. C. and Suzuki, K. (1971). G_{M1}-gangliosidosis. Correlation of clinical and biochemical data. *Arch. Neurol.,* **24,** 58

Svennerholm, L. (1966). The metabolism of gangliosides in cerebral lipidoses. *In: Inborn Errors of Sphingolipid Metabolism.* (New York: Delfgauw Press)

Swallow, D. M., Stokes, D. C., Corney, G. and Harris, H. (1974). Differences between the hexosaminidase isoenzymes in serum and tissues, *Ann. Hum. Genet.,* **37,** 287

Thomas, G. H., Taylor, H. A., Miller, C. S., Axelman, J. and Migeon, B. R. (1974). Genetic complementation after fusion of Tay–Sachs and Sandhoff cells. *Nature (London),* **250,** 580

van Someren, H. and Beijersbergen van Henegouwen, H. (1973). Independent loss of human Hex A and Hex B in man–Chinese hamster somatic cell hybrids. *Humangenetik,* **18,** 171

Weerd-Kastelein, E. A. de, Keijzer, W. and Bootsma, D. (1972). Genetic heterogeneity of xeroderma pigmentosum demonstrated by somatic cell hybridization. *Nature (London),* **238,** 80

Wolfe, L. S., Callahan, J., Fawcett, J. S., Andermann, F. and Scriver, C. R. (1970). G_{M1}-gangliosidosis without chondrodystrophy and visceromegaly. *Neurology,* **20,** 23

Yamamoto, A., Adachi, S., Kamamura, S., Takahashi, M., Kitani, T., Ohtori, T., Shinji, Y. and Nishikawa, M. (1974). Localized β-galactosidase deficiency. *Arch. Int. Med.,* **134,** 627

12
The Cell Membrane in Metabolic Control: The Regulation of Cholesterol Biosynthesis in Familial Hypercholesterolaemia

F. W. ROBERTSON

Department of Genetics, University of Aberdeen

The increasing incidence of coronary heart disease has stimulated research into causes across a broad front. Some of the most promising studies have been concerned with biochemical investigation into cholesterol synthesis and its regulation and also genetic investigations of the incidence of coronary heart disease, especially in relation to levels of lipoprotein. It is convenient to consider the main features of the genetic evidence first.

Familial hypercholesterolaemia has long been recognized as a monogenic effect leading in homozygotes to extremely high levels of serum cholesterol, xanthomata, early development of atherosclerosis and a low chance of survival beyond 30 years of age. Heterozygotes show cholesterol levels roughly intermediate between the homozygotes and normals, so the phenotype is compatible with more or less additive effect of the alleles concerned although the mutant is persistently and wrongly identified as dominant in the literature. While no one doubts the existence of a monogenic determination of this form of elevated cholesterol, there is at present disagreement as to how far either monogenic or polygenic variation contributes generally to the elevated levels of cholesterol so often associated with coronary heart disease, a controversy which embraces also the origin of elevated levels of triglyceride.

The case for a major role of monogenic determination has been provided in a massive study of the familial incidence of total serum cholesterol and total triglyceride concentrations by a group of workers in Seattle (Goldstein *et al.*, 1973a, 1973b). These authors put forward the hypothesis, based on pedigree data and the occurrence of bimodal distributions in relatives of survivors of myocardial infarction, that either elevated serum cholesterol alone or elevated serum triglyceride alone or joint elevation of both lipids, are under separate monogenic control, while the contribution to coronary heart disease by polygenic variation of serum lipids is comparatively minor. It is difficult however to arrive at unambiguous genetic interpretation of continuously variable traits. Apparent segregations pose problems as to the criteria of abnormality. Although it is usual to regard a value at or above the 90th or 95th percentile as 'abnormal' for routine diagnostic purposes, there is plenty of scope for differing opinions as to whether given differences reflect the effect of segregation at only a few or many loci.

In hypercholesterolaemia (Goldstein *et al.*, 1973a, 1973b) a strong case has been made for the importance of segregation at one or two loci from the clear evidence of bimodality in cholesterol levels in relatives of index cases compared with the unimodality of controls. Similar evidence has been obtained in our Aberdeen study of serum lipoprotein values in the families of survivors of myocardial infarction living in north-east Scotland.

The need to arrive at a clear understanding of the origin of elevated serum cholesterol is indicated by the evidence that in the American study some 10% of myocardial infarction survivors showed abnormally high values and in our study, which refers to survivors under 50 years of age, the figure works out at about 20%. Obviously if monogenic segregation contributes appreciably to these totals the need for unambiguous identification of index cases and possible screening of their relatives calls for consideration.

THE DEVELOPMENT OF A HYPOTHESIS BY BROWN AND GOLDSTEIN (1976) FOR THE RECEPTOR MEDIATED CONTROL OF CHOLESTEROL METABOLISM

One of the principal collaborators in this family study, J. L. Goldstein, moved on to investigate, with M. S. Brown, the biochemistry of the genetic effect by the use of skin fibroblasts taken from normal individuals and from persons who were inferred to be homozygous or heterozygous for the hypercholesterolaemia gene on the basis of pedigree and clinical evidence. The important initial finding was that fibroblasts from normal individuals and from homozygotes show characteristic differences in cholesterol synthesis although the interpretation of how this behaviour related to atherosclerosis is still a matter for debate. Figure 12.1 summarizes the sequence of cholesterol mechanisms proposed by Brown and Goldstein (1976a).

Following the lead provided by Bailey's (1967) demonstration that cultured mouse cells show an increased ability to synthesize cholesterol when incubated

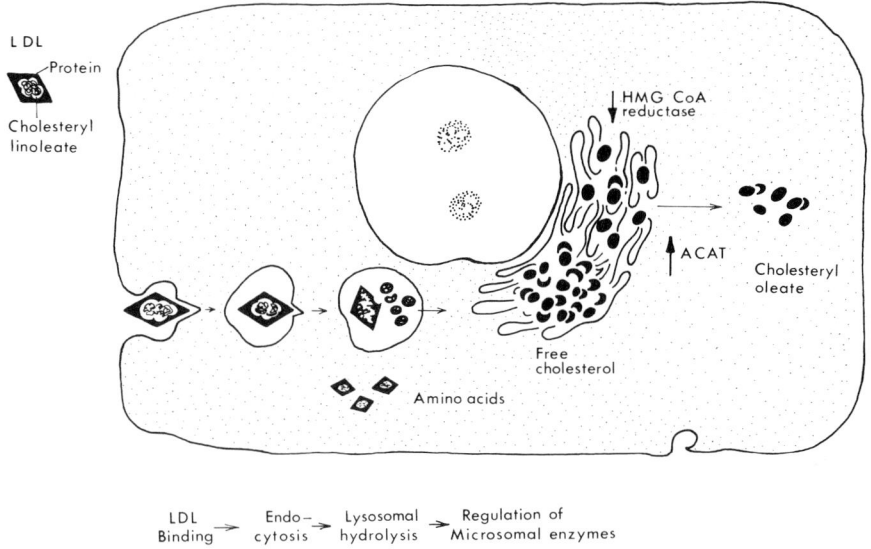

Figure 12.1 The sequence of LDL metabolism in cultured fibroblasts. (From Brown and Goldstein, 1976a).

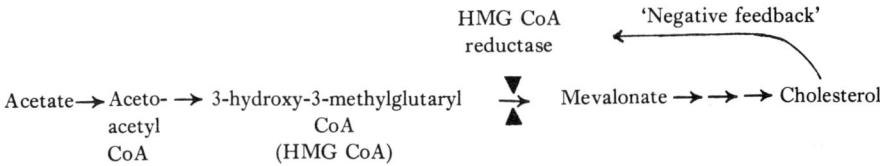

Figure 12.2 Summary of pathway of synthesis of cholesterol from acetate in mammalian liver.

in a cholesterol-free medium, Goldstein and Brown have focused attention on the regulation of the rate-limiting enzyme in cholesterol synthesis, 3-hydroxy-3-methylglutaryl coenzyme A (HMG CoA) reductase, which occupies the position shown in Figure 12.2, which summarizes the pathway of cholesterol synthesis in liver. Sipersteen and Guest (1960) and Sipersteen and Fagan (1966) showed that high dietary cholesterol intakes were accompanied by a reduced capacity of the liver to convert acetate to cholesterol, without reduction in the ability to convert acetate to HMG CoA nor of ability to convert mevalolactone to cholesterol, thereby identifying the conversion of HMG CoA to mevalolactone as the step involved in feedback inhibition. Following this lead, experiments were set up to study the regulation of the reductase in skin fibroblasts from contrasted genotypes. The cells were grown in monolayer for five to fifteen generations before use generally in Eagle's essential medium (MEM), supplemented with non-essential amino acids, and in the presence of

10% fetal calf serum. HMG CoA reductase activity was measured in terms of the formation of [3-^{14}C] mevalonate from [3-^{14}C]hydroxymethylglutaryl CoA, using thin-layer chromatography to separate the mevalolactone.

When cells were transferred from such a medium containing fetal calf serum to a lipid-free medium there was a striking increase in enzyme activity (Figure 12.3). When cells have been incubated in MEM without serum for a period long

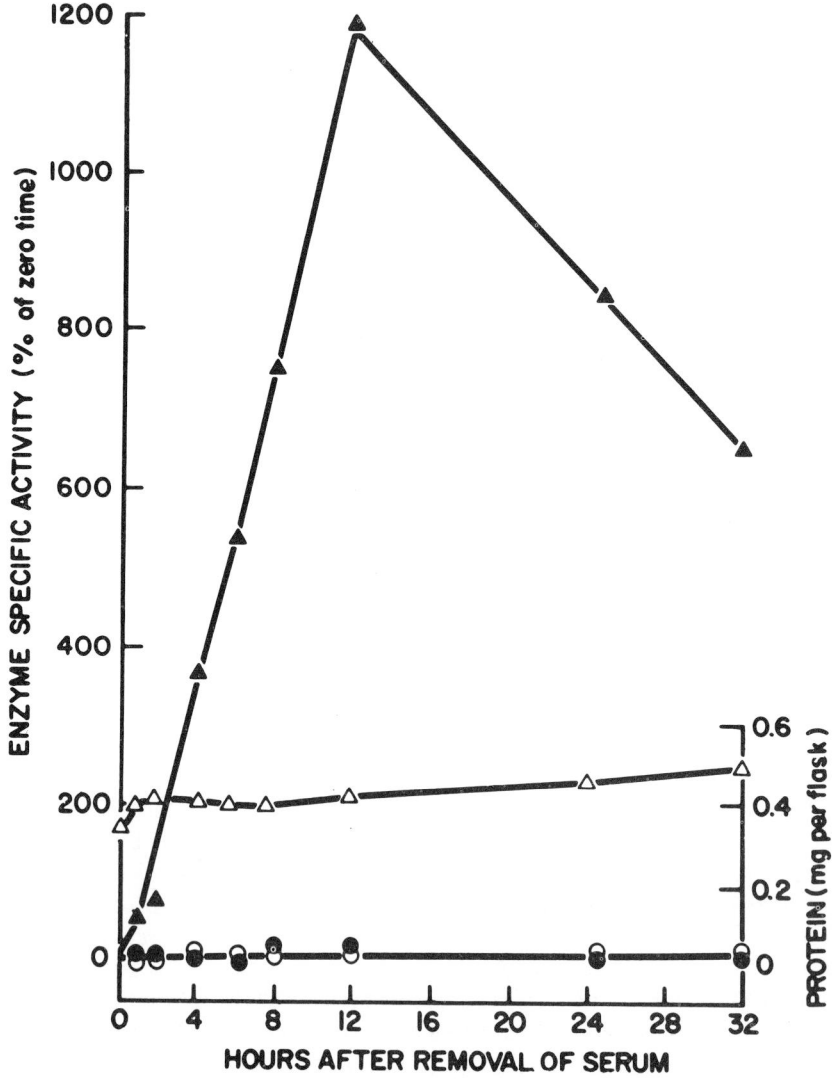

Figure 12.3   Increase in HMG CoA reductase in skin fibroblasts after removal from fetal calf serum to a lipid-free medium: △ . protein content. ▲ : HMG CoA reductase activity. ● : lactate dehydrogenase activity. ○ : acid phosphatase activity. (From Brown *et al.*, 1973b)

enough to lead to high enzyme activity, transfer to medium with serum led to a sharp decline in activity. The presence of serum from different mammalian species produced similar effects. Lipoproteins were identified as the relevant factors in serum (Figure 12.4a, 12.4b, Figure 12.5) and this was confirmed by

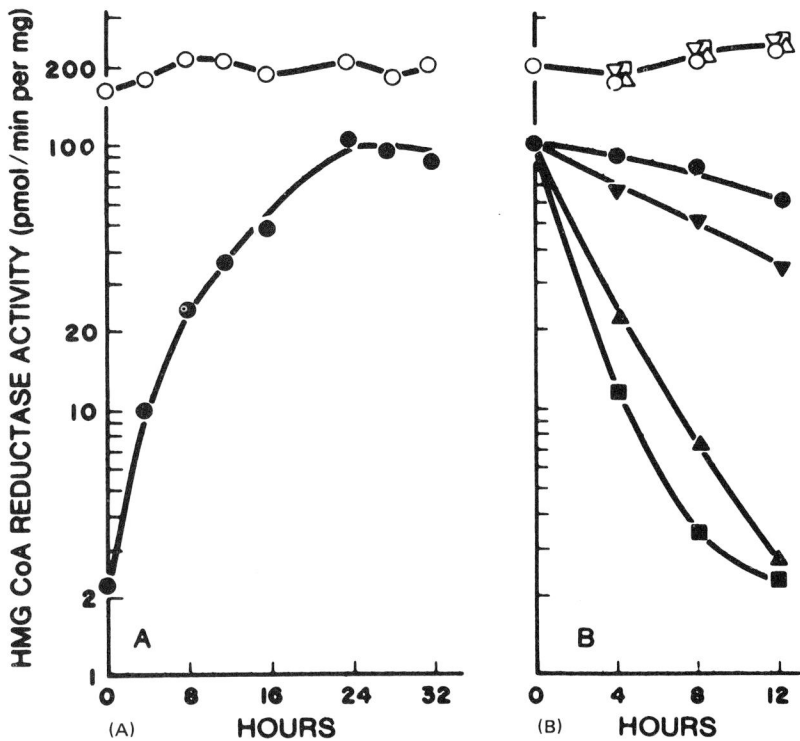

Figure 12.4 HMG CoA reductase activity in fibroblasts of a control subject ($\bigcirc \triangledown \triangle \square$) and of a person homozygous for hypercholesterolaemia ($\bullet \blacktriangledown \blacktriangle \blacksquare$). (A) After six days' growth in medium containing 10% calf serum, the medium was replaced with fresh medium containing 5% human lipoprotein-deficient plasma. Extracts were prepared at intervals. (B) 24 h after addition of the lipoprotein-deficient plasma, buffer containing alternative amounts of human LDL were added to provide different final concentrations. Concentration of LDL culture medium: $\bigcirc \bullet$ none; $\triangledown \blacktriangledown$ 2 μg/ml; $\triangle \blacktriangle$ 10 μg/ml; $\square \blacksquare$ 20 μg/ml. (From Goldstein and Brown, 1973)

testing the effects of ultracentrifugally separated preparations of low-density lipoprotein (LDL) and high-density lipoprotein (HDL). HDL proved ineffective, although in the published figures there is at least some evidence of a decline. The similar action of very low density lipoproteins (VLDL) and LDL was attributed to the presence of apolipoprotein B which is present in both these lipoproteins but is absent in HDL (Brown *et al.*, 1974).

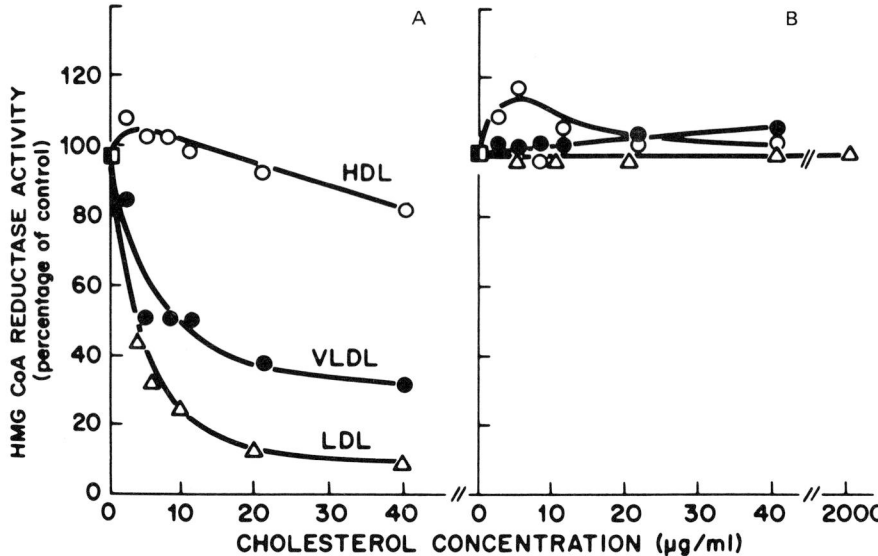

Figure 12.5   Effect of addition to the medium of alternative lipoproteins on HMG CoA reductase activity. (A) Normal fibroblasts. (B) Fibroblasts from a hypercholesterolemic patient. (Symbols for the different lipoproteins in B are the same as in A.) (From Brown *et al.*, 1974)

The lowering of reductase activity in the presence of serum containing lipid was noted only when the fibroblasts were intact; in cell extracts there was no effect. Also when extracts from treated cells with high reductase activity were combined with extracts from cells with low activity, intermediate amounts of mevalonate were produced, suggesting that reduced activity was not due to an intracellular inhibitor of the enzyme. Since both cycloheximide and actinomycin D were effective in shutting off activity it seemed likely that the increase in enzyme activity required *de novo* protein synthesis.

When similar tests were applied to fibroblasts from an individual homozygous for the hypercholesterolaemia gene it was shown (Figure 12.4) that such cells differed from control cells in having a very high level of activity in the presence of fetal calf serum (Goldstein and Brown, 1973) and addition of either of the main lipoprotein types HDL, LDL, VLDL in place of whole serum had no effect (Figure 12.5). This contrast between genotypes in enzyme activity according to the presence or absence of lipid carrying serum in the incubation medium is the primary observation which various investigators are seeking to interpret. Further tests showed that the LDL from a hypercholesterolaemic subject did not differ from normal LDL in its regulation of HMG CoA reductase activity in normal cells (Figure 12.6).

When enzyme activity in the presence of differing amounts of LDL was compared in cells from homozygous individuals, from normals and from 'heterozygous' individuals, the heterozygotes were found to be roughly in-

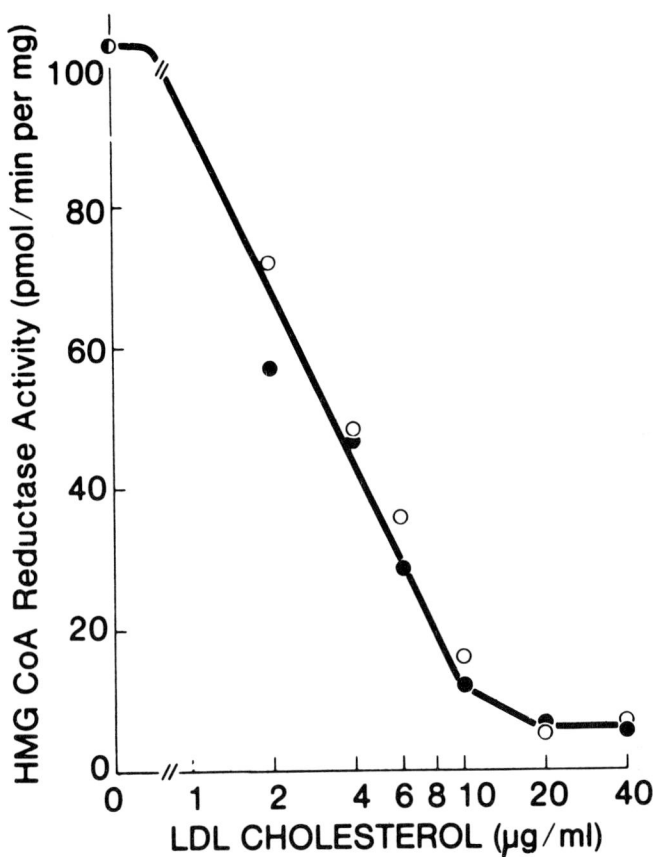

Figure 12.6   Effect of LDL from a control (●) or a hypercholesterolemic homozygote (○) in suppressing HMG CoA reductase in normal cells. (From Goldstein and Brown, 1973)

termediate in the lowering of activity associated with increased amounts of LDL (Figure 12.7).

In view of the absence of evidence for allosteric effects on the enzyme and also the similar kinetic properties of HMG CoA reductase from normal individuals and from homozygotes, it was inferred that the genetically controlled lesion lies not in the enzyme itself but in some other protein which is involved in the regulatory process (Goldstein and Brown, 1973).

To confirm the role of LDL and VLDL in the regulation process Brown, Dana and Goldstein (1974) studied the effects of treating high activity cells with serum from a patient suffering from abetalipoproteinaemia, in which apolipoprotein B is absent. As anticipated such serum failed to cause a lowering of HMG CoA reductase. When fibroblasts from normal individuals or from a homozygote were grown for 24 h in a medium containing non-lipoprotein cholesterol the reductase activity in both types was reduced by more than 70%

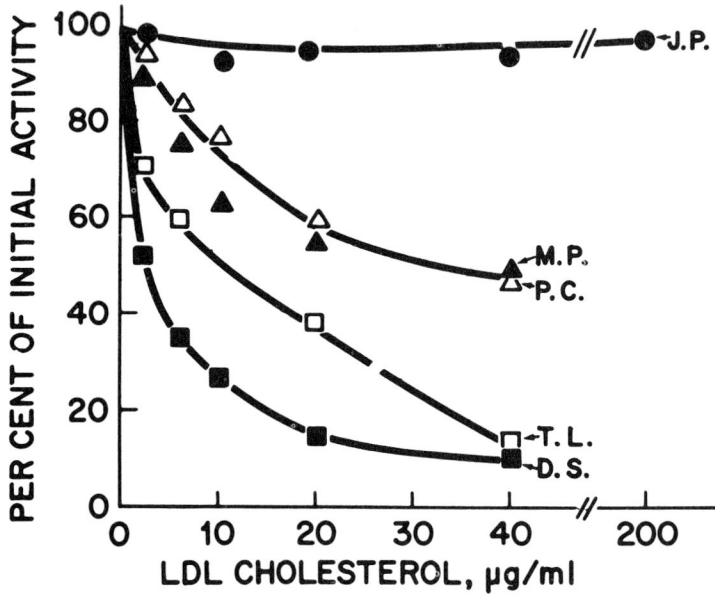

Figure 12.7   Effect of LDL on HMG CoA reductase activity in skin fibroblasts from control subjects (□■); heterozygotes (△▲) and a person homozygous for hypercholesterolemia (●). (From Goldstein and Brown, 1973)

suggesting that cholesterol is the ultimate suppressor of activity but that the circumstances of its binding in the form of LDL or VLDL determines whether or not it is to be effective in influencing HMG CoA reductase activity.

At this stage in their work Brown and his colleagues (1974) inferred that the LDL cholesterol reduces reductase activity by inhibiting the synthesis of the enzyme. The measurement of enzyme turnover and the exponential decline in activity after addition of cycloheximide indicated a short half-life of about three hours in cultured cells, similar to that reported for the enzyme in rat liver (Edwards and Gould, 1972). As noted earlier, experiments with cell extracts do not provide evidence of direct inhibition of the enzyme.

The next major piece of evidence (Goldstein and Brown, 1974) concerned the measurement of binding of LDL to the fibroblasts by the use of [^{125}I]LDL prepared with iodine monochloride (MacFarlane, 1958). Control tests showed that such labelled LDL behaved exactly as normal LDL in reducing HMG CoA reductase activity. By incubating cells in the presence of labelled LDL for various periods and at different concentrations it was possible to follow the course of binding in cells from normal individuals and from homozygotes (Figure 12.8). Specific binding, defined as a difference between the label bound

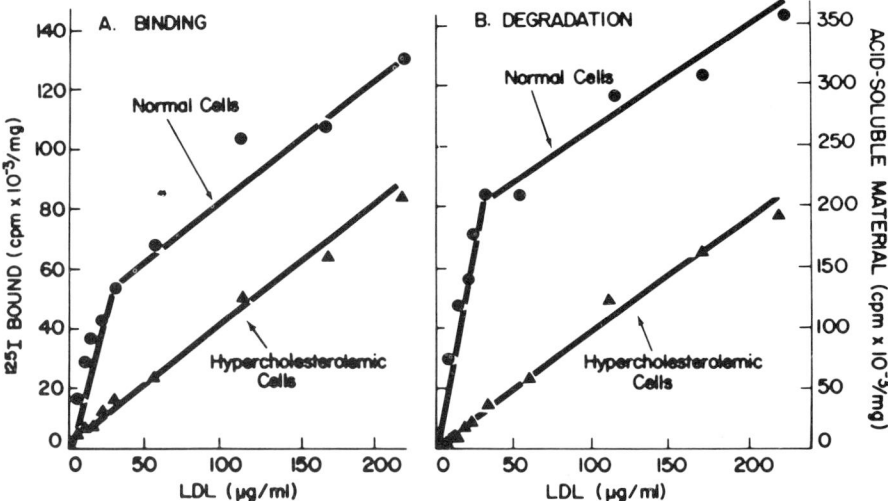

Figure 12.8 The extent of binding and rate of degradation in normal and hypercholesterolemic cells with varying concentrations of labelled LDL. (A) Shows the difference in high affinity binding compared with similar rates of unspecific binding, while (B) shows the rate of degredation of bound LDL after incubation at 37 °C for 6 h. (From Goldstein and Brown, 1974)

in the absence or presence of excess LDL, reached a plateau at about three hours. Specific binding of label was virtually nil in fibroblasts from a homozygous hypercholesterolaemic individual.

In normal cells at a concentration of LDL below 25 μg/ml the amount of binding rose sharply and linearly with increasing amount of LDL (Figure 12.8). This form of binding was designated as high affinity binding and it showed saturation. At levels of LDL above 25 μg/ml binding continued to occur with increasing concentration of LDL but the rate was much less as indicated by the shallow slope in Figure 12.8. Whereas cells from hypercholesterolaemic people showed no high affinity binding, both normal and mutant cells showed the same rate of non-specific binding. Also, when the rate of degradation of labelled LDL was determined after 6 hours' incubation at 37 °C corresponding estimates of the rates of degradation and binding tallied exactly. At levels of LDL less than 15 μg/ml the ratio of specific to non-specific binding was 10:1. Theoretical analysis of the data indicated the presence of specific binding sites of high affinity and it was estimated that at 37°C a maximum of 250,000 molecules of LDL per cell can be bound. At 4 °C there was a similar reaction although the maximum number of molecules which could be bound was estimated at only 7,500 per cell. After disruption of fibroblasts, incubated at either

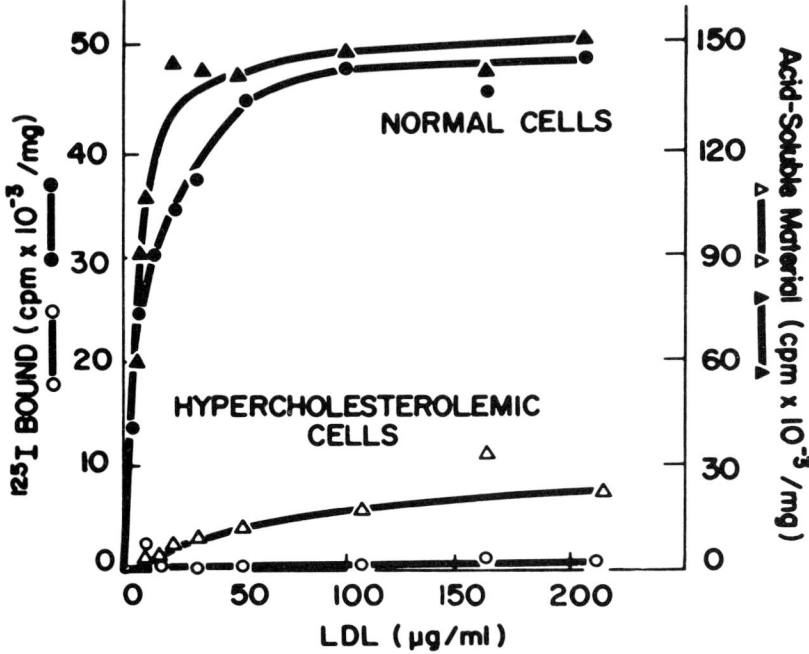

Figure 12.9   High affinity binding (O ●) and degradation (△▲) of labelled LDL in normal and hyperlipidemic cells was measured by subtracting the counts per minute bound or degraded in the presence of native LDL from that recorded in its absence. (From Goldstein and Brown, 1974)

temperature with labelled LDL, 85% of the bound label sedimented by centrifugation at 10,000 g for 5 min suggesting that the labelled LDL was bound to the cell membrane.

Goldstein and Brown (1974) next considered whether cell membrane receptors were directly concerned in the specific binding or whether apparently bound material had been transferred to the cell interior. They showed that when fibroblasts are incubated with labelled LDL, the lipoprotein is degraded by a first order process which is temperature dependent. Most of the released label is dialysable and less than 25% can be precipitated by 10% trichloracetic acid (Figures 12.8 and 12.9).

Evidence in favour of high affinity binding sites on the cell membrane was obtained by treatment of cells with pronase under conditions known to alter the composition of the cell surface (Burger, 1970). With this pronase treatment (Figure 12.10) there was a 66% reduction in high affinity binding and a 55%

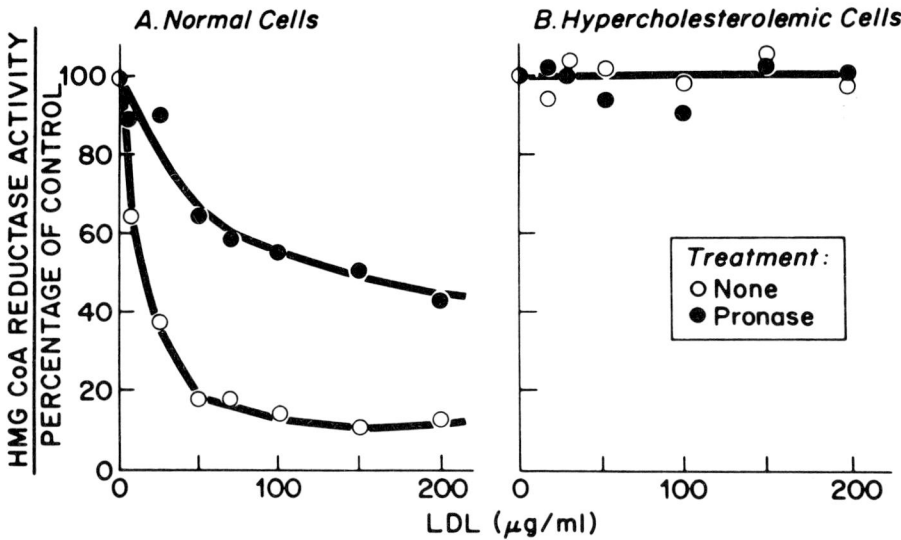

Figure 12.10   Effect of pronase treatment of cells on the ability to react with LDL by suppression of HMG CoA reductase. (From Goldstein and Brown, 1974)

reduction in the rate of LDL degradation in normal cells. Since neither binding nor degradation was significantly reduced by this treatment in the mutant cells it was inferred that the so-called high affinity sites were chiefly involved. The unspecific low affinity binding was unaffected by pronase treatment.

At this stage in the study the authors supported the hypothesis of specific binding of LDL at high affinity sites on the cell surface. The suppression of HMG CoA reductase activity and degradation of the LDL to a dialysable form were part of a coordinated process. The lesion in the homozygous cholesterolaemic patients appeared to be a defect in the binding reaction. Unspecific or low affinity binding was interpreted in terms of unspecific endocytosis at similar rates in both normal and mutant cells. Although LDL is degraded to a dialysable form, whatever its method of cell entry, there remained the problem that although LDL suppression of HMG CoA reductase results from the release of cholesterol via degradation of the lipoprotein, apparently similar degradation of LDL absorbed by the low affinity process has no effect in suppressing HMG CoA reductase in cells from homozygous hypercholesterolaemic individuals.

Further possibilities were examined in a study of the esterification of LDL cholesterol in cultured fibroblasts from normal and homozygous hypercholesterolaemic individuals (Goldstein *et al.*, 1974). Fibroblasts were grown in lipid-deficient medium and the cells were transferred to medium containing 5% lipid-deficient serum with either an appropriate concentration of LDL or free cholesterol. After incubation either [1-^{14}C]acetate or [1-^{14}C]oleate–albumen was added and the cells incubated at 37 °C for a period after which the labelled cholesterol and cholesteryl esters were recovered. As anticipated, cells in the presence of increasing amounts of LDL lead to a progressive decline of incorporation into labelled cholesterol due to reduction in the activity of HMG CoA reductase. At levels of LDL at 100 μg/ml or above, the incorporation became barely detectable. On the other hand, the incorporation of labelled acetate into cholesteryl esters, hardly detectable in the absence of LDL, was greatly increased in its presence (Table 12.1). After alkaline

Table 12.1   Incorporation of labelled acetate into lipids of cultured cells from normal individuals. (After Goldstein *et al.*, 1974)

	[^{14}C]acetate incorporation into lipids (*pmoles/Dish*)				
Addition to incubation medium	[^{14}C]- fatty acids	[^{14}C]- triglycerides	[^{14}C]- phospholipids	*Free* [^{14}C]- cholesterol	*Cholesteryl* [^{14}C]- esters
None	165	32	414	700	18
LDL	177	34	461	32	198
HDL	125	30	409	635	13

hydrolysis of the labelled cholesteryl esters almost all the counts were located in the fatty acid fraction. In sharp contrast, when cells from the homozygous hypercholesterolaemic individual were treated in the same way, the addition of LDL had no effect on the incorporation of [^{14}C]acetate. Thus, measurement of esterification with labelled acetate provided a further criterion for characterizing the primary effect of the hypercholesterolaemia gene.

When the fibroblasts were supplied with free cholesterol in the medium the rate of esterification in normal cells was notably higher when cholesterol was bound in the form of LDL than in the free state but the situation was reversed for the mutant cells which are thus able to esterify cholesterol.

Brown and Goldstein (1974) have used evidence from comparisons between cells from 13 different normal individuals and also cells from five homozygote and 11 heterozygote individuals to support their interpretation of receptor-mediated control of LDL incorporation into cells. It is worth noting that six of the heterozygotes were parents of homozygotes and hence there was little doubt about their genotype. Using the criteria of binding of labelled LDL and its degradation as well as the record of reduction of HMG CoA reductase activity, they found that in heterozygotes the level of binding was reduced to about 40% below the normal level. This was associated with, a decline of about 40% in the level of LDL degradation and with approximately half the normal level of suppression of HMG CoA reductase. The relationship between LDL degradation and LDL binding conformed to a single regression slope for the three genotypes (Figure 12.11). Homozygotes showed no significant reduction

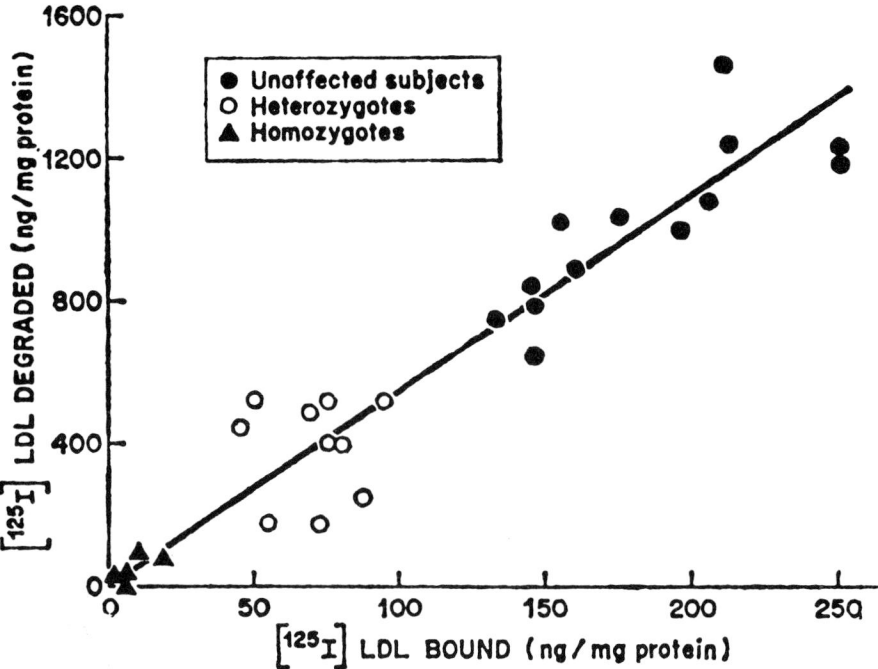

Figure 12.11   The relations between the binding and degradation of LDL in cells from normal individuals, heterozygotes and homozygotes for the hypercholesterolemia gene. (From Brown and Goldstein, 1974)

of HMG CoA reductase at concentrations of LDL which were 50–90% higher than levels which lead to 90% suppression in normal cells. They argued that if, *in vivo*, the hypothetical receptor molecules are chiefly involved in regulating cholesterol synthesis and LDL degradation and if the processes are controlled at LDL levels which involve only a proportion of the total number of receptors, then, although heterozygotes would resemble normal individuals in absolute rates of cholesterol synthesis and LDL degradation, a 2.5 fold higher concentration of LDL would be needed in heterozygotes to effect an equivalent occupation of receptor sites and hence an equivalent repression of HMG CoA reductase activity.

Brown, Dana and Goldstein (1975) investigated the fate of bound LDL by following the consequence of its hydrolysis. LDL was labelled by incubation of tritiated cholesteryl linoleate with LDL in the presence of 10% dimethylsulphoxide. Incorporation of the label into LDL led to no change in the properties in terms of suppression of HMG CoA reductase and activation of cholesteryl ester formation in the fibroblasts. When such labelled LDL was incubated with fibroblasts the resulting tritiated cholesterol accumulated in the cell. The rate of hydrolysis was greater in normal than in cells from homozygous hypercholesterolaemic individuals. Addition of chloroquine which blocks degradation processes in lysosomes of cultured cells (de Duve et al., 1974) blocked the cholesterol ester hydrolysis. In mutant cells the rate of hydrolysis was very low. However, cell-free extracts from both normal and mutant fibroblasts behaved alike in effectively hydrolysing the LDL-bound labelled linoleate cholesterol ester at similar rates. There was also clear evidence of re-esterification of free labelled cholesterol obtained from the LDL as measured by [^{14}C]oleate incorporation. On such evidence the authors concluded that the LDL bound to receptors becomes incorporated in lysosomes in which the cholesteryl esters and the protein components are hydrolysed to release free cholesterol which is transferred to membranes. Cholesterol synthesis is suppressed by inhibition of HMG CoA reductase while cholesteryl ester synthesis is activated by stimulation of cholesteryl acyltransferase.

Brown and colleagues (1975) further showed that incubation of fibroblasts with certain oxygenated sterols, namely 25-hydroxycholesterol, 7-ketocholesterol or 6-ketocholesterol, sharply increased the rate of esterification of endogenous cholesterol and led to an increase in the cell content of cholesterol esters, apparently due to a correlated increase in activity of fatty acyl CoA: cholesterylacyltransferase. Since the same sterols also caused suppression of HMG CoA reductase activity, they inferred that cholesteryl ester formation and cholesterol synthesis in fibroblasts are jointly regulated in a reciprocal manner.

The role of lysosomes in the metabolism of LDL was studied in fibroblasts from a patient with a genetically determined cholesteryl ester storage disease in which lysosomal acid lipase activity was deficient. The mutant cells behaved normally with respect to high affinity binding of LDL and absorption of the lipoprotein into the cell interior. The defect in lysosomal hydrolysis of the cholesteryl ester moiety of the lipoprotein caused the accumulation of cholesteryl esters. On this evidence it was inferred that the LDL is hydrolysed within lysosomes with a consequent release of free cholesterol.

The impressive evidence accumulated by these investigators (summarized in Figure 12.12) has provided the basis for a general theory of the regulation of cholesterol content of fibroblasts illustrated in Figure 12.13 which is reproduced from the paper by Brown and Goldstein (1976a). In human fibroblasts lack of cholesterol leads to an increase in the synthesis of LDL receptor molecules which are localized on the cell surface. The first step in LDL metabolism is the binding of LDL to a receptor, a reaction which shows saturation, high affinity and specificity, in that only lipoproteins containing apolipoprotein B are bound. An ionic interaction may be involved in the binding, since heparin which forms a complex with LDL through ionic interac-

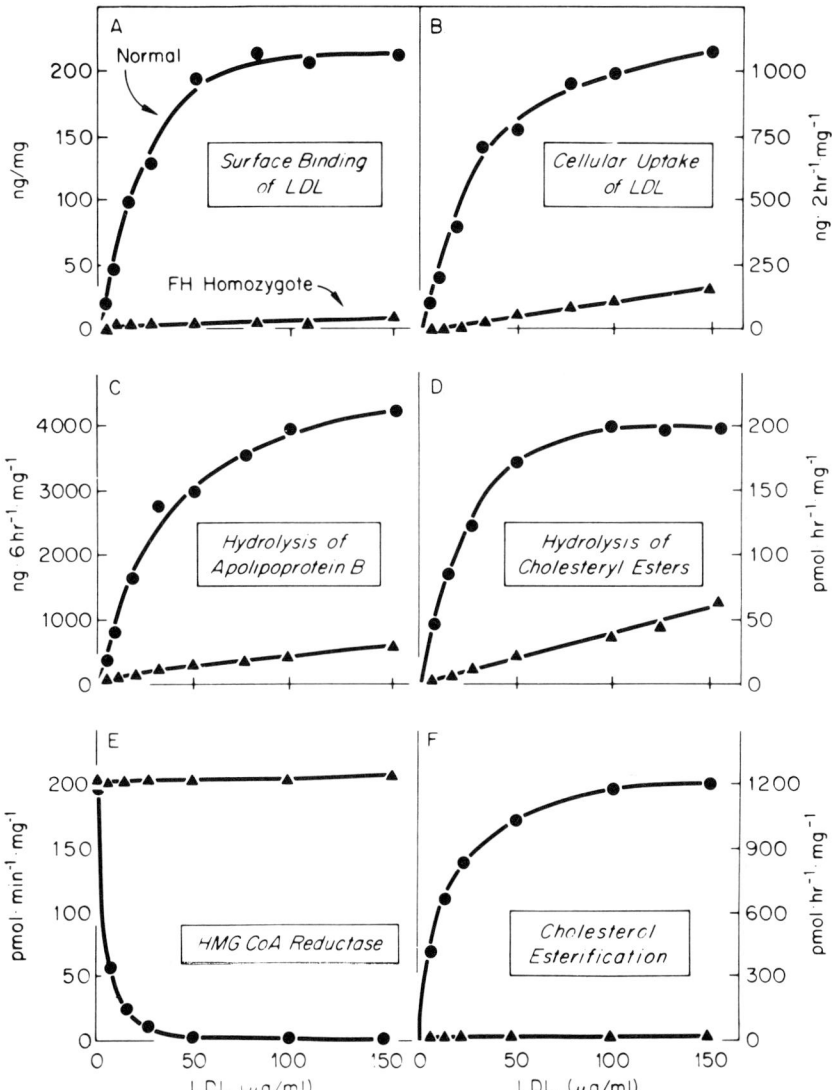

Figure 12.12 Summary of proposed coordinated sequence of events which start with the binding of LDL to fibroblasts in cells from normal individuals and from individuals homozygous for the hypercholesterolemia gene. (From Brown and Goldstein, 1976a)

tion, can prevent binding of LDL and also cause dissociation. The bound LDL enters the cell by an active process which resembles 'absorptive endocytosis'. Absorbed LDL is incorporated into endosomes which fuse with lysosomes in which the LDL protein is degraded to amino acids which are released into the medium. The cholesteryl ester component of LDL is hydrolysed by a lysosomal

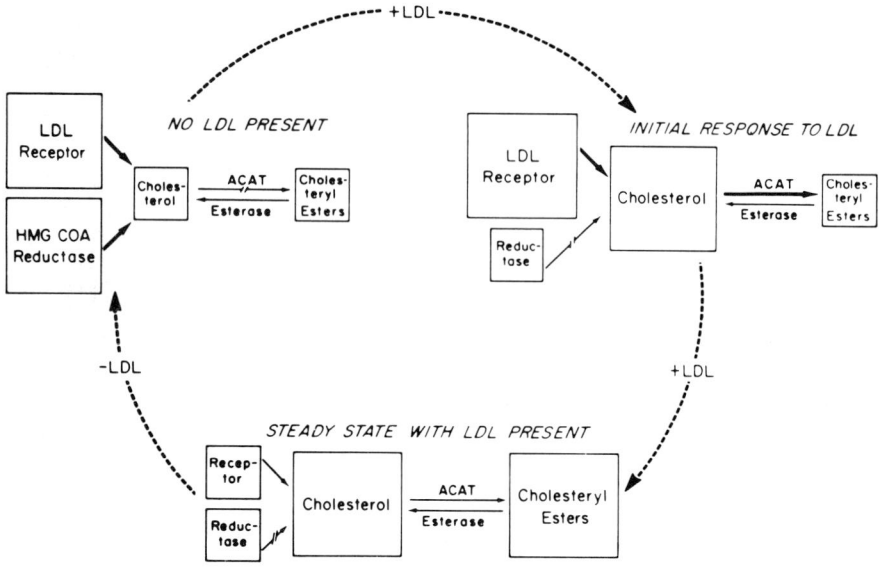

Figure 12.13 Regulation of the cholesterol content of cultured human fibroblasts by LDL according to the hypothesis developed by Brown and Goldstein. The relative amount of each component is indicated by the size of the square. (ACAT is fatty acyl CoA: cholesterylacyltransferase.) (From Brown and Goldstein, 1976a)

acid lipase, the unesterified cholesterol becoming largely associated with cell membranes. Accumulation of unesterified cholesterol suppresses HMG CoA reductase, thereby reducing cholesterol synthesis, and also activating an acyl-CoA: cholesteryl acyltransferase which results in its re-esterification. This re-esterified cholesterol is then preferentially attached to the mono-unsaturated fatty acids, oleate and palmitate in contrast to the cholesteryl esters of plasma LDL which are rich in the polyunsaturated fatty acid, linoleate. In this scheme (Figure 12.13) the LDL receptor plays a major role in transfer of free and esterified cholesterol from LDL into the cell and is also responsible for a change in the fatty acid composition of cholesteryl esters from a polyunsaturated to a more saturated form. In the usual steady state in the presence of whole serum, cholesterol synthesis is suppressed and cells preferentially bind and use of LDL derived cholesterol while the receptor activity or the numbers of receptors are adjusted to maintain a constant sterol concentration in the cell. Under atypical conditions in which the LDL concentration in the medium is abnormally low or when there is a genetically controlled deficiency of receptor molecules, cholesterol synthesis is increased as part of a homeostatic mechanism to maintain cell growth and division. This interpretation is regarded as the key to the earlier observations of Bailey (1967) that, in the presence of serum, cholesterol synthesis is minimal in mammalian cells which thus preferentially use exogenous cholesterol.

Monogenic hypercholesteraemia is accounted for within this framework since in fibroblasts derived from affected individuals, LDL binding, hydrolysis of apolipoprotein B, cholesteryl ester hydrolysis, suppression of HMG CoA reductase and activation of cholesteryl ester formation are all greatly reduced in homozygotes due to their presumed lack of the key LDL receptor. The intermediate activity shown by heterozygotes is what might be expected in the presence of half the usual number of LDL receptors.

An additional route of cholesterol into the cell is via unspecific ingestion so that lipoprotein is taken in at a rate proportional to its concentration outside the cell. LDL absorbed by this route is apparently incorporated in lysosomes where its protein and cholesteryl esters are subject to hydrolysis but there is apparently a difference in that the free cholesterol derived from LDL by the passive route is apparently excreted into the culture medium and neither suppresses HMG CoA reductase nor activates cholesteryl ester formation in normal or mutant cells.

These studies have to be related to events in the body. Brown and Goldstein (1976a), in discussing the occurrence of atherosclerosis in patients with normal plasma lipoprotein values, have argued that, if the LDL receptors behave in the same way in the body as they do in fibroblasts, their role in cholesteryl ester deposition would be different in individuals who develop atherosclerosis in the presence or absence of abnormally elevated concentrations of plasma LDL. In genetically determined hypercholesterolaemia, defective receptors lead to high plasma levels because of a restriction of degradation of LDL and also a restriction of HMG CoA reductase repression but in patients, with normal LDL and cholesterol levels, the LDL receptor might also have a localized effect on the arterial wall.

The authors also draw a distinction between hepatic and non-hepatic tissues and suggest that in tissues other than liver cholesterol synthesis is suppressed because the tissues preferentially absorb and use LDL through the receptor route. On the other hand, in the liver, which along with the intestine, is responsible for more than 90% of the cholesterol produced in the body, they suggest that cholesterol synthesis is unlikely to be regulated by the LDL receptor; this is supported by the primary influence of dietary cholesterol on cholesterol syntheses.

The major features of the receptor hypothesis are summarized in Figure 12.11. Let us now consider some evidence which either complicates the genetic interpretation or which even sheds doubt on the receptor theory itself.

EVIDENCE AGAINST THE HYPOTHESIS OF RECEPTOR-MEDIATED CONTROL OF CHOLESTEROL METABOLISM

Dealing first with the genetic aspects, there is evidence from several sources that genetically determined hypercholesterolaemia is not a single entity. Thus the Dallas group themselves reported that among patients who were rated as homozygous, some showed the extreme phenotype in which the receptors appeared to be absent while others showed intermediate responses. This was interpreted as evidence for reduction rather than absence of receptors (Goldstein and Brown, 1976a). The two types have been referred to as respectively 'receptor negative' and 'receptor defective'. Although it was impossible to distinguish the alternative genotypes with confidence by measurement of LDL binding, such a discrimination was possible in terms of cholesteryl ester formation in the presence of LDL. Breslow et al. (1975) have reported patients with homozygous familial hypercholesterolaemia which differ in their response to diet and drug therapy. Fibroblasts were derived from each type and were compared with respect to the suppression of HMG CoA reductase by LDL and also binding of labelled LDL. In cells from patients who did not respond to diet or drugs the enzyme was not inhibited by increased levels of LDL, whereas cells from those who did respond to treatment showed partial inhibition. All patients showed defective binding of LDL but this was less marked in the patients responding to drug or diet treatment. The authors suggested that these differences might be due to such genetic alternatives as one locus with at least three alleles or two separate loci with two or more alleles per locus. Avigan Bhathena and Schreiner (1975) studied HMG CoA reductase activity in fibroblasts from four unrelated, homozygous hypercholesterolaemic individuals. Their cells in clear contrast with typical receptor negative cells showed suppression of HMG CoA reductase activity in the presence of 10% whole serum. The authors stress the possible genetic heterogeneity of the hypercholesterolaemic condition. Betteridge, Higgins and Gatton (1975) have reported a rather different situation in which the genetically determined lesion in a monogenically determined form of hypercholesterolaemia appears to be located in the low density lipoprotein molecule itself. The individuals concerned have an abnormal LDL which does not suppress the activity of HMG CoA reductase in their own leucocytes or in leucocytes of unrelated controls. If lipoprotein is derived from individuals unrelated to the patient, LDL asserts its usual effect and suppresses the enzyme in the hypercholesterolaemic individuals.

Stein and co-workers (1976) have attacked the concept of LDL binding to specific receptors as the primary step in LDL degradation. Using fibroblasts from a person with homozygous hypercholesterolaemia and fibroblasts from control individuals, the LDL binding was followed by use of [^{125}I]LDL. The amount transferred to the interior of the cell (cell entry) was determined after 4–6 minutes' incubation of the labelled cells in 0.05% trypsin. This released a proportion of the label into the medium while the rest of the label, which is not released, is regarded as having been transferred into the interior of the cell. General handling of cells was similar to that used by the Goldstein and Brown group.

Figure 12.4 shows the time curve for binding of labelled LDL, cell entry and production of TCA soluble material in normal fibroblasts. Binding was initially

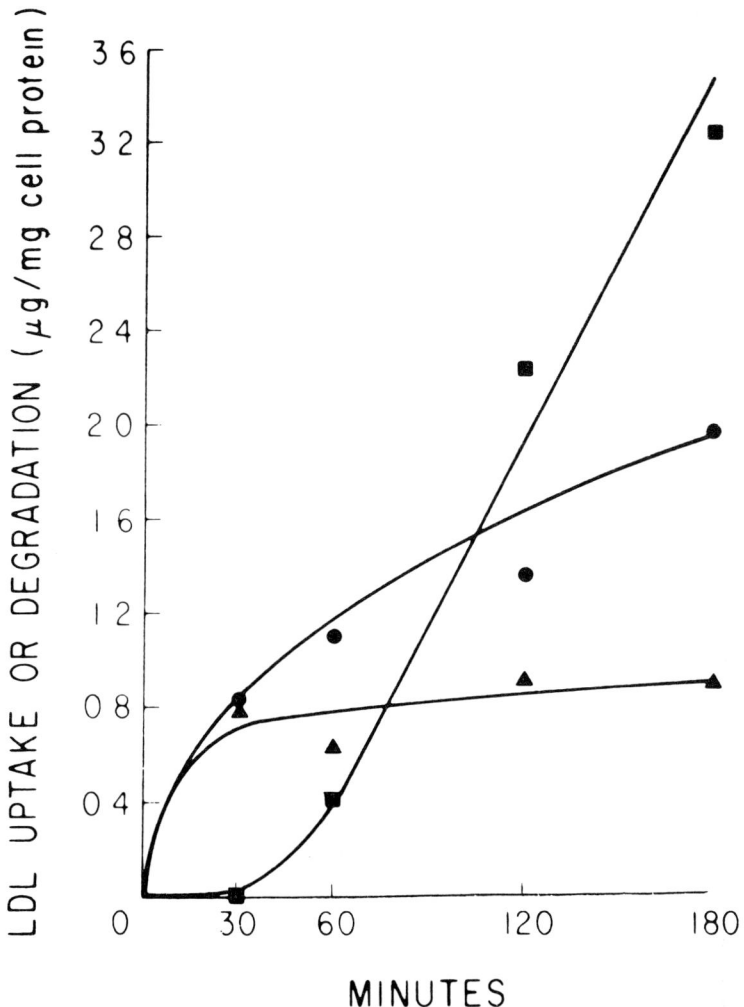

Figure 12.14   Time course for binding of labelled LDL (▲), cell entry (●) and production of trichloracetic acid soluble material (■) by normal fibroblasts. (From Stein *et al.*, 1976)

fast and reached a maximum after about 30 min. The transfer into the cell equalled or exceeded the binding level over short periods and soon became substantially greater. Cells from normal and homozygous individuals were compared at 0 and 37 °C respectively (Figure 12.15). At 0 °C binding proceeded in both cell types at a rate proportional to LDL concentration although not in a

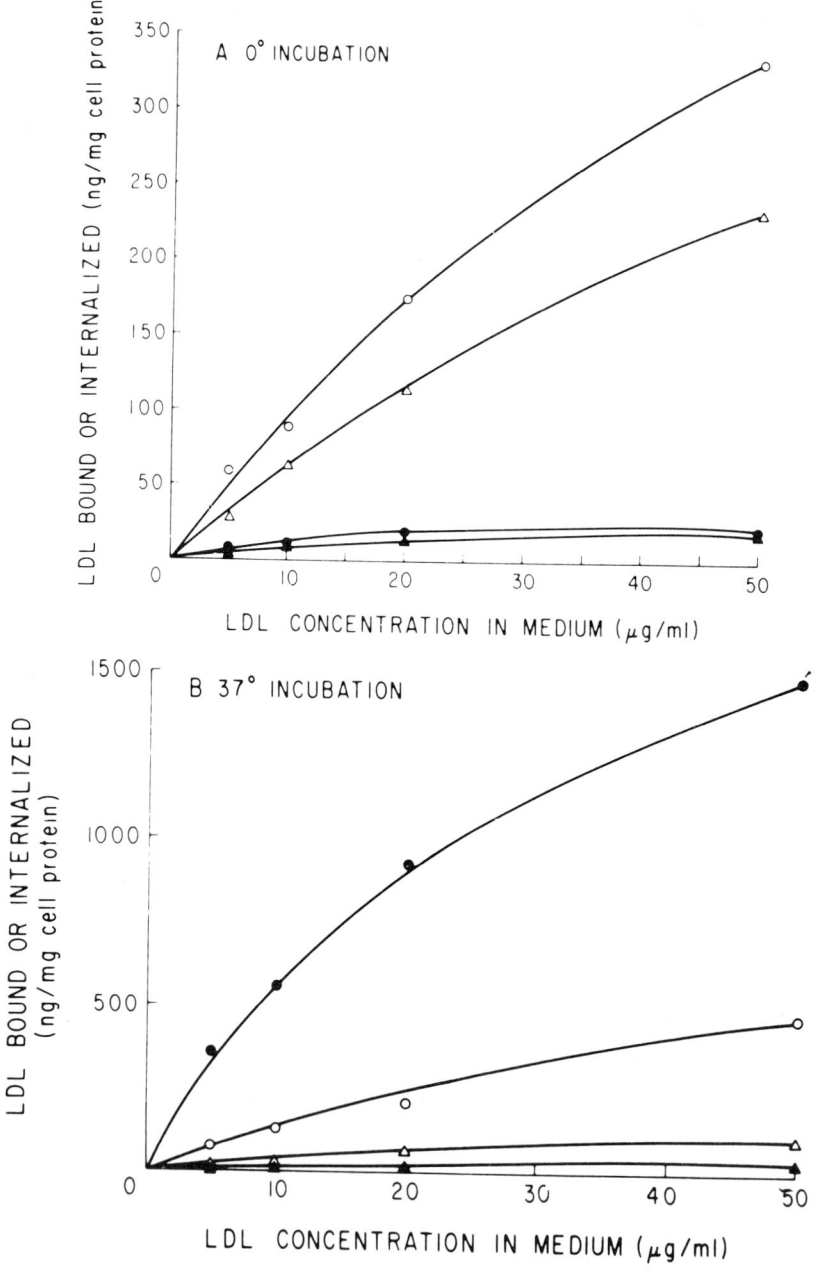

Figure 12.15 Comparison of binding (O△) and cell entry (●▲) by fibroblasts from normal individuals (●▲) and from a patient homozygous for hypercholesterolemia (△▲). (A) at 0°C incubation. (B) at 37°C incubation. (From Stein *et al.*, 1976)

linear fashion. Normal cells bound a relatively greater amount of LDL but the difference was small and cell entry was extremely slow in both cell types. At 37 °C cell entry (internalization) greatly increased in normal cells but showed hardly any change in the cells from the homozygotes. The ratio of trypsin-released label at 37 °C worked out at an average of 1.12 and 0.52 for the homozygotes, so there is apparently a clear difference in temperature influenced binding.

Comparison of binding at 0 °C over a wide range of concentrations showed that above a concentration of 100 μg LDL per ml there was little difference in the level of binding between the mutant and normal cell types. At low LDL levels binding to normal cells was greater but the curve relating binding and concentration apparently crosses at 60 μg LDL/ml. At 0 °C there is no detectable degradation of labelled LDL. At 37 °C degradation in mutant cells reached only 20–47% of the level attained by normal cells. If it is accepted that the sum of LDL internalized and the LDL degraded approximates to the total taken into the cell, the genotypes may be compared for relative rates of degradation. Over a 3–6 h period approximately 50% of the internalized LDL is degraded in normal cells compared with about 90% by mutant cells. Apparently therefore, mutant cells are fully capable of LDL degradation if the lipoprotein has been taken up by the cell.

Stein and colleagues (1976) compare their evidence with that of Goldstein and Brown (1976a) and conclude that the difference between normal and mutant fibroblasts resides in the difference in the process of internalization rather than in the presence or absence of binding to the surface. Their data show a 50–100 fold difference in the former and a comparatively minor difference in the latter. However, Brown and Goldstein (1976b) have challenged this evidence and attribute the results of Stein and colleagues (1976) to an artefact which makes it often difficult to distinguish between specific binding to receptors and specific binding to other components in the culture.

A rather different and more fundamental criticism of the Goldstein and Brown approach has been advanced by Fogelman and colleagues (1975) who have studied HMG CoA reductase activity in leucocytes from heterozygotes for the hypercholesterolaemia gene and also from normal individuals. Sterol synthesis from [^{14}C]acetate was linearly proportional to the number of leucocytes and highly reproducible. Leucocytes grown for 6 hours on lipid free medium containing labelled acetate produced more sterol from this source than when incubated in whole serum but the composition of the medium had no effect when sterol was synthesized from mevalonate. Since acetate and mevalonate were present in excess, this confirms for man that the rate-limiting step in cholesterol synthesis precedes mevalonate formation.

Leucocytes from heterozygotes and controls were incubated with [^{14}C]acetate in whole and also lipid-free serum. In whole serum no significant difference between normal leucocytes and cells from heterozygous individuals was noted. In the lipid-free serum, however, incorporation by heterozygote cells was approximately three times greater compared with only a two-fold increase by control cells. Identical results were obtained when the cells were incubated in lipid-free sera with either 0.44 or 0.88 mM acetate.

The acetate incorporated into sterol in lipid-free medium minus the amount incorporated into sterol in whole sera was attributed to 'induced enzyme'. Comparison of heterozygotes and normals in the presence of excess acetate showed about twice as large an activation of acetate utilization in heterozygotes as in normal cells. This is in agreement with earlier evidence from similar experiments in which heterozygotes were induced to a higher level of activity in lipid-free serum. Such differences are highly correlated with levels of HMG CoA reductase.

The amounts of mevalonate formed by HMG CoA reductase and the amount of mevalonate formed from [^{14}C]acetate to account for [^{14}C] incorporation into sterols were compared. There is such a close correspondence that the synthesis of [^{14}C]sterols from [^{14}C]acetate may be taken as a measure of HMG CoA reductase activity.

To compare normal and heterozygous cells with respect to acetate incorporation into sterols and induction of HMG CoA reductase, the time course of the induction of the enzyme was determined by measuring enzyme activity in freshly isolated cells and also after different periods of incubation in lipid-free serum. In freshly isolated cells reductase activities were similar in the two types but thereafter the heterozygote cells showed an increased rate of HMG CoA reductase activation.

The induction of the enzyme in lipid-free serum and the higher activity in heterozygous cells call for an explanation. Since the lipid-free medium contains the apoproteins of the lipoproteins, stripped of their sterol and triglyceride moieties, the possibilities are that enzyme induction represents a response to a normal loss of sterols in cells and also that heterozygotes may lose cellular sterols faster than the controls. Cells from both types of individual were grown for three hours in whole serum containing tritiated mevalonate. After checking the level of label in a small sample of cells, the rest were washed and transferred to lipid-free sera. At intervals the tritiated sterol contents of the medium and of the cells were measured. The cell pellets and the media were also analysed for tritiated squalene after each period of incubation. There was no evidence of any deterioration of the cells. After the first three hours in whole serum, normal cells had lost 13% of total labelled sterols into the medium while the heterozygotes had lost 21%. After transfer to lipid-free medium both normal and heterozygous cells lost sterol into the medium but the former soon reached a comparatively steady rate while the heterozygous cells continued to lose labelled sterol. After 16 hours, 40% of the total sterol had been lost from the heterozygous cells compared with only 26% from the normals. The continued loss of sterol from heterozygous cells was in contrast with the small loss of tritiated squalene which was also monitored after each transfer to lipid-free medium. In both types of cell the intracellular squalene content remained unchanged so that it appears there is a relatively much larger loss of sterol from the leucocytes and also that this loss is much larger from heterozygous than from normal cells.

In comparing their evidence with that of Goldstein and Brown, Fogelman and colleagues (1975) recognize the current impossibility of direct comparison since LDL binding has apparently not yet been studied in leucocytes although both sets of observations agree that the genetic abnormality in hypercholesterolaemia is concerned with synthesis of HMG CoA reductase. The interpretation of Fogelman and colleagues (1975) is fundamentally different since they believe that the abnormality involves defective binding for a sterol repressor which is synthesized within the cell and with a consequently greater loss of sterol from the cell. Given a sufficiently high extracellular level of cholesterol, rates of cholesterol synthesis and loss of sterol are restored to normal values.

A problem which begs explanation is posed by the now famous case of the 12 year old girl (J. P.) homozygous for familial hypercholesterolaemia whose cells were used by Goldstein and Brown (1973) in the early fibroblast studies. After a myocadial infarction her serum cholesterol was initially reduced by intravenous feeding. A portacaval shunt was performed and her serum cholesterol concentration fell dramatically. It is not obvious how this case history can be reconciled with a genetically determined lesion in binding of LDL to cells. Fogelman and colleagues (1975) draw attention to the hazards of argument by analogy from the behaviour of fibroblasts or leucocytes *in vitro* to what goes on in the body. They consider that fibroblasts which are grown in lipid-free medium are subject to complete derepression of HMG CoA reductase which can never occur *in vivo*. Such apparently conflicting evidence from fibroblasts and leucocytes along with the strong indications of genetic heterogeneity in hypercholesterolaemia together with our current ignorance of the degree of association between genetically determined differences in plasma cholesterol levels and isolated cell behaviour suggest that we have quite a long way to go before there is general agreement about the origin of hypercholesterolaemia.

In concluding this survey we cannot do better than draw attention to the concluding comments of Fogelman and his colleagues (1975): 'A complete understanding of the origin of familial hypercholesterolaemia has not yet been achieved. In *vivo* studies have so far failed to show an over-production of cholesterol while *in vitro* tests for HMG CoA reductase activity certainly provide no evidence for higher levels in heterozygotes'. Few would dispute that fibroblasts and/or leucocytes will provide valuable diagnostic tools for distinguishing between genotypes and will eventually lead to a more complete understanding of familial hypercholesterolaemia.

References

Avigan, J., Bhathena, S. J. and Schreiner, M. E. (1975). Control of sterol synthesis and of hydroxymethylglutaryl CoA reductase in skin fibroblasts grown from patients with homozygous type II hyperlipoproteinemia. *J. Lipid Res.*, **16**, 151

Bailey, J. M. (1967). *In*: G. H. Rothblat and D. Kritchevsky (eds.) p. 85, *Lipid Metabolism in Tissue Culture Cells*. (Philadelphia: Wistar Institute Press)

Betteridge, D. J., Higgins, M. J. P. and Galton, D. J. (1975). Regulation of 3-hydroxy-3-methylglutaryl coenzyme-A reductase activity in type II hyperlipoproteinaemia. *Br. Med. J.*, **4**, 500

Breslow, J. L., Spaulding, D. R., Lux, S. F., Levy, R. I. and Less, R. S. (1975). Homozygous familial hypercholesterolemia. *N. Engl. J. Med.*, **293**, 900

Brown, M. S., Goldstein, J. L. and Siperstein, M. D. (1973a). Regulation of cholesterol synthesis in normal and malignant tissue. *Fed. Proc. Fed. Am. Soc. Exp. Biol.*, **32**, 2168

Brown, M. S., Dana, S. E. and Goldstein, J. L. (1973b). Regulation of 3-hydroxy-3-methylglutaryl coenzyme A reductase activity in human fibroblasts by lipoproteins. *Proc. Nat. Acad. Sci. U.S.A.*, **70** (7), 2162

Brown, M. S., Dana, S. E. and Goldstein, J. L. (1974). Regulation of 3-hydroxy-3-methylglutaryl coenzyme A reductase activity in cultured human fibroblasts. *J. Biol. Chem.*, **249**, 789

Brown, M. S. and Goldstein, J. L. (1974). Expression of the familial hypercholesterolemia gene in heterozygotes: mechanism for a dominant disorder of man. *Science*, **185**, 61

Brown, M. S., Dana, S. E. and Goldstein, J. L. (1975). Cholesterol ester formation in cultured human fibroblasts. Stimulation by oxygenated sterols. *J. Biol. Chem.*, **250**, 4025

Brown, M. S. and Goldstein, J. L. (1976). Receptor-mediated control of cholesterol metabolism. *Science*, **191**, 150

Brown, M. S. and Goldstein, J. L. (1976). Familial hypercholesterolemia: a genetic defect in the low-density lipoprotein receptor. *New Eng. Jour. Med.*, **294**, 1386

Burger, M. M. (1970). Proteolytic enzymes initiating cell division and escape from contact inhibition of growth. *Nature*, **227**, 170

de Duve, C., de Barsy, T., Poole, B., Trouet, A., Tulkens, P. and van Hoof, F. (1974). Lysosomatic agents. *Biochem. Pharmacol.*, **23**, 2495

Edwards, P. A. and Gould, R. G. (1972). Turnover rate of hepatic 3-hydroxy-3-methylglutaryl coenzyme A reductase as determined by use of cycloheximide. *J. Biol. Chem.*, **247**, 1520

Fogelman, A. M., Edmond, J., Seager, J. and Popjak, G. (1975). Abnormal induction of 3-hydroxy-3-methylglutaryl coenzyme A reductase in leukocytes from subjects with heterozygous familial hypercholesterolemia. *J. Biol. Chem.*, **250**, 2045

Goldstein, J. L. and Brown, M. S. (1973). Familial hypercholesterolemia: identification of defect in the regulation of 3-hydroxy-3-methylglutaryl coenzyme A reductase activity associated with overproduction of cholesterol. *Proc. Nat. Acad. Sci. U.S.A.*, **70**, 2804

Goldstein, J. L., Hazzard, W. R. and Schrott, H. G. (1973a). Hyperlipidemia in coronary heart disease I. Lipid levels in 500 survivors of myocardial infarction. *J. Clin. Invest.*, **52**, 1533

Goldstein, J. L., Hazzard, W. R. and Schrott, H. G. (1973b). Hyperlipidemia in coronary heart disease. II. Genetic analysis of lipid levels in 176 families and delineation of a new inherited disorder combined hyperlipidemia. *J. Clin. Invest.*, **52**, 1544

Goldstein, J. L. and Brown, M. S. (1974). Binding and degradation of low density lipoproteins by cultured human fibroblasts. *J. Biol. Chem.*, **249**, 5153

Goldstein, J., Dana, S. E. and Brown, M. S. (1974). Esterification of low density lipoprotein cholesterol in human fibroblasts and its absence in homozygous familial hypercholesterolemia. *Proc. Nat. Acad. Sci. U.S.A.*, **71**, 4288

McFarlane, A. S. (1958). Efficient trace-labelling of proteins with iodine. *Nature* (*London*), **182**, 53

Siperstein, M. D. and Guest, M. J. (1960). Studies on the feed-back control of cholesterol synthesis. *J. Clin. Invest.*, **39**, 642

Siperstein, M. D. and Fagan, J. M. (1966). Feed-back control of mevalonate synthesis by dietary cholesterol. *J. Biol. Chem.*, **241**, 602

Stein, O., Weinstein, D. B., Stein, Y. and Steinberg, D. (1976). Binding, internationalization and degradation of low density lipoprotein by normal human fibroblasts and by fibroblasts from a case of homozygous familial hypercholesterdemia. *Proc. Nat. Acad. Sci. U.S.A.*, **73**, 14

Milner Lecture

13

Newer Developments in Tissue Culture: Further Aid in the Study of Inborn Errors of Metabolism

J. E. SEEGMILLER

University of California, San Diego, U.S.A.

The surge of progress in recent years in the recognition and characterization of human inborn errors of metabolism has far surpassed the most optimistic expectations of those of us who began working in this field as recently as 25 years ago. The magnitude of this change is illustrated by a comparison of the historical development of just two diseases. The classical inborn error of metabolism, alcaptonuria, was first described clinically over 300 years ago (LaDu, 1972). Some 250 years elapsed before the concept of the metabolic defect was first proposed. Sir Archibald Garrod formulated the gene–enzyme relationship with a defective gene giving rise to an inborn error of metabolism in his Croonian Lecture of 1908 which thereby establishes him as the scientific father of the society meeting here today. Fully another half-century elapsed before his hypothesis was proven by actual demonstration of a deficiency of the enzyme homogentisic acid oxidase in the liver of an affected patient (LaDu *et al.*, 1958). By contrast only three years elapsed between the clinical description of the Lesch–Nyhan syndrome in 1964 (Lesch and Nyhan, 1964) and the demonstration of the deficient enzyme of purine metabolism responsible for this devastating neurological disease (Seegmiller *et al.*, 1967).

When one considers the series of unlikely events that were needed to permit a meaningful study of almost any inborn error of metabolism during those 50 years following Garrod's contribution, progress was achieved against overwhelming odds. In the case of alcaptonuria, it required the bringing together of a patient with this rare disease and physicians with the appropriate training in biochemistry. This concurrence was greatly aided by the establishment of the Clinical Center of our National Institutes of Health where this work was performed. But this alone was not enough. On the first encounter the physicians felt that they could not morally justify the liver biopsy that would be required since the information to be obtained would not change the management of the patient and therefore could not contribute immediately to the patient's welfare. Only with the patient's subsequent admission with massive gastrointestinal bleeding requiring abdominal surgery was an incidental biopsy taken of the liver which permitted the demonstration of the isolated deficiency of this single enzyme of tyrosine metabolism (LaDu et al., 1958).

A major factor contributing to the more rapid progress made in recent years has been the use of tissue culture. Human fibroblasts cultured from virtually innocuous skin biopsies have provided a readily available source of patients' cells for characterization of the abnormal gene products responsible for many inborn errors of metabolism. In a recent compilation of known enzyme defects, around two-thirds were expressed in fibroblasts (Raivio and Seegmiller, 1972). Accessibility of these mutant cells to a far wider range of investigators has been aided substantially by the establishment of a central repository for the storage and disbursement of strains of well-characterized mutant human cells obtained from patients with an ever increasing number of different hereditary diseases.* Such a development now allows biochemists without a medical degree to make contributions often of a sophisticated nature. More recently the type of cells available has been expanded to include permanent cell lines of human lymphoblasts that can be readily established with a very high success rate from 10 ml or even smaller samples of peripheral blood (Sly et al., 1976). The present state of muscle and nerve cell culture has been reviewed in this volume by Dubowitz and by Thompson. For those disorders in which the affected gene is expressed only in hepatic, renal or other specialized cells of the body satisfactory means have yet to be devised for their detailed investigation in cultured cells.

Nevertheless, substantial progress has been made. Over the past 20 years the number of enzyme defects identified as causes of hereditary diseases have increased at a logarithmic rate with a doubling time of around four years (McKusick, 1975). The majority of these diseases have been relatively rare recessively inherited disorders. Nevertheless, they have contributed substantially to our knowledge of genetic mechanisms of disease. In recent years considerable evidence has accumulated of hereditary factors, many of them dominantly inherited, contributing to many of our major diseases causing death and disability. Included are diabetes mellitus, the cardiovascular diseases of arteriosclerosis, myocardial infarction and hypertension, the affective disorders of manic depressive psychosis and

---

* Information on the Human Genetic Mutant Cell Repository can be obtained by writing to Dr Arthur Greene, Institute for Medical Research, Copewood and Davis Streets, Camden, New Jersey, U.S.A. The repository is sponsored by the National Institute of General Medical Sciences.

schizophrenia, many forms of neuromuscular disease, an increasing number of disorders of the immune system and certain forms of arthritis. The task of extending the identification of abnormal gene products to include these more common diseases is a most challenging one for the years ahead. Only recently has a start been made with the identification of a deficiency of a receptor protein in cultured fibroblasts as the cause of one form of familial hypercholesterolemia (Goldstein and Brown, 1974) showing a dominant inheritance pattern and a number of enzyme defects responsible for gouty arthritis, one of which is dominantly inherited (Becker and Seegmiller, 1975).

Despite this surge of progress we have made only a start in the understanding of the human genome. The quantity of DNA per human cell is sufficient to code for 10^7 polypeptides of 150 amino acids each. However, much of the DNA is reiterated sequences. Only about 1% of the total amount would be adequate to account for the estimated number of human genes (Table 13.1). Yet we have

Table 13.1   Current status of knowledge of the human genome

Modality measured or calculated	Number observed or estimated
Potential number of polypeptides of 150 amino acids calculated from DNA per cell	10^7
Estimated number of human genes	10^5
Well-substantiated human hereditary disorders	10^3
Abnormal gene products identified	2×10^2

From McKusick, 1975

clinical evidence so far of only about 10^3 well-substantiated human hereditary disorders (McKusick, 1975) and of those we have identified the abnormal gene product in less than 200 genetic aberrations. Furthermore, over 95% of those identified are for recessively inherited disorders. We have just begun to identify the gene products responsible for dominantly inherited disorders yet they account for more than half of clinically recognized genetic traits (Table 13.2). Presumably these abnormalities are produced by defects in structural proteins, transport mechanisms and proteins of the cell wall in addition to abnormalities of receptor proteins and enzymes of high specific activity (Becker et al., 1973) already demonstrated.

THE INVESTIGATION OF PURINE METABOLISM WITH THE AID OF CULTURED SKIN FIBROBLASTS

In the 15 years since cultured human cells were first introduced for the study of inborn errors of metabolism by Krooth and Weinberg (1961), cultured fibroblasts have been used not only to document specific enzyme defects but also to obtain new insight into the mechanisms by which the enzyme defect produces the clinical features of the specific disease. In my own laboratory we have had a long-standing interest in purine metabolism and its regulation. Detailed studies of fibroblasts cultured from normal subjects and from patients who produce excessive quantities of uric acid have substantially expanded our knowledge of this subject as well as providing insight into the role of purine metabolism in the normal functioning of other metabolic processes.

The most extreme example of excessive production of purines in the human species was first described by Lesch and Nyhan in 1964 in two young brothers, one of whom presented with haematuria and a marked uric acid crystalluria (Lesch and Nyhan, 1964). Both brothers produced around six times the normal amount of uric acid in the 24 h urine. In addition, they showed a severe neurological disease consisting of choreoathetosis, spasticity, a moderate degree of mental retardation and a most bizarre compulsive self-mutilation by biting away the lips and tongue as well as the ends of the fingers. In retrospect, a report by Riley in 1960 of a child in Scotland with severe juvenile gout was probably the same disorder since the characteristic severe self-mutilation was apparent in a photograph taken at autopsy (Boyle and Buchanan, 1971).

As an index of the rate of purine synthesis in cultured fibroblasts we have added the glutamine analogue azaserine which blocks purine biosynthesis *de novo* at a stage just three steps removed from the presumed rate-determining reaction catalysed by the enzyme phosphoribosylpyrophosphate glutamine amidotransferase (PAT). As a result of the azaserine block, the substrate formylglycineamide ribonucleotide (FGAR) accumulates. The incorporation of [^{14}C]formate into the FGAR then provides an index of the rate of these reactions. Fibroblasts cultured from affected children showed a three to five-fold increase in [^{14}C]formate incorporation in this sytem, a value quite comparable to the degree of excessive uric acid synthesis shown by the children *in vivo* (Seegmiller *et al.*, 1967; Rosenbloom *et al.*, 1968). Additional similarities between the *in vitro* cell culture system and studies *in vivo* were found. Azathioprine failed to decrease the rate of uric acid synthesis of affected children *in vivo* while it produced a substantial inhibition of uric acid synthesis when administered to gouty volunteers (Kelley *et al.*, 1968). Azathioprine is metabolized *in vivo* to 6-mercaptopurine which inhibits the synthesis *de novo* of purines by normal human fibroblasts but fails to inhibit purine synthesis in fibroblasts from children with the Lesch–Nyhan syndrome. With the subsequent demonstration of a gross deficiency of the enzyme hypoxanthine–guanine phosphoribosyltransferase (HPRT) in all tissues of affected children, this difference in response of affected children was readily explainable since the deficient enzyme is required for conversion of 6-mercaptopurine to its metabolically active nucleotide form.

The HPRT enzyme defect allowed us to ask and to answer a very simple question. Is the action of the HPRT enzyme required for the immunosuppressive action of azathioprine? With the help of Dr Thomas Waldmann of our National Cancer Institute we investigated various aspects of the immune response in affected children (Seegmiller *et al.*, 1977). In contrast to the report of Allison and colleagues (1975) we found normal concentrations of the various classes of gamma-globulins in the plasma of four affected children. As might be expected, lymphocytes isolated from their blood showed no inhibition of either their proliferation or of their antibody production in response to mitogenic stimulation at concentrations of 6-mercaptopurine that completely inhibited the response of normal lymphocytes. On the other hand 6-methylmercapto-purine ribonucleoside which requires a different enzyme for conversion to the biologically active nucleotide form produced a similar inhibition in both normal and HPRT deficient lymphocytes (Seegmiller et al., 1977).

Study of cultured fibroblasts has given us new insight into the cause of the overproduction of purines observed in affected children. A feedback inhibition

FEEDBACK CONTROL OF PURINE SYNTHESIS

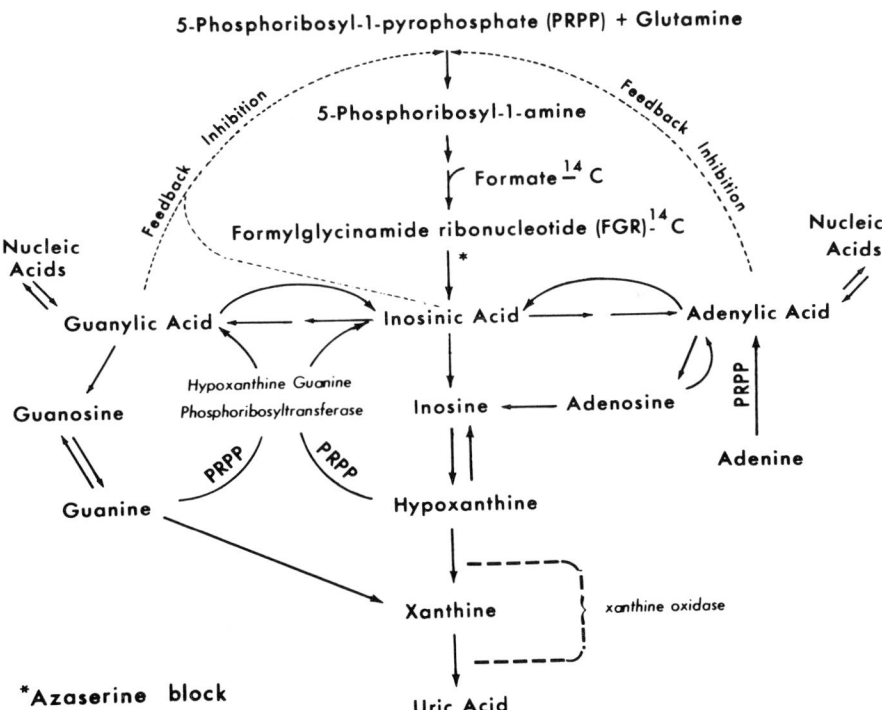

Figure 13.1 Regulation of purine synthesis and site of the enzyme defect in the Lesch–Nyhan syndrome. Although purine nucleotides inhibit the presumed rate-limiting enzyme phosphoribosylpyrophosphate (PP-ribose-P) glutamine amidotransferase, the excessive rate of purine synthesis in the Lesch–Nyhan syndrome is apparently driven by the accumulation of PP-ribose-P, a substrate for the missing enzyme as well as for the rate-limiting amidotransferase. (Reproduced from Seegmiller et al., Science, 155, 1682, 1967, with permission of the publisher. Copyright 1967 by the American Association for the Advancement of Science)

of the rate-controlling PAT enzyme of purine synthesis by purine nucleotides has long been postulated (Figure 13.1). A lowering of the intracellular concentration of purine nucleotides as a result of the HPRT deficiency could very well remove such an inhibition and thereby result in an acceleration of the rate of purine synthesis by converting the PAT enzyme from its less active aggregated form to a more active disaggregated form (Figure 13.2) as proposed by Holmes

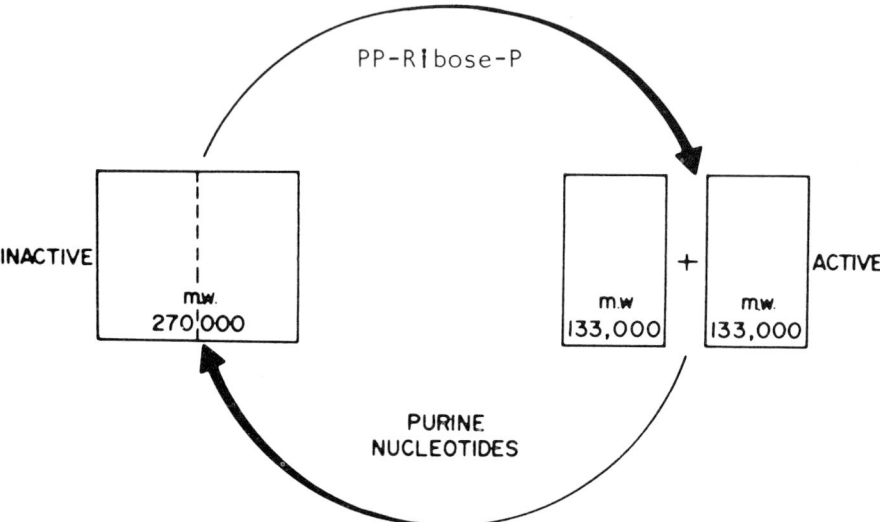

Figure 13.2   The molecular mechanism of regulation of the rate of purine biosynthesis. The activity of the enzyme phosphoribosylpyrophosphate (PP-ribose-P) glutamine amidotransferase is regulated by its conversion to a relatively inactive aggregated form by purine nucleotides and its reversal to an active disaggregated form by PP-ribose-P. (Reproduced from Holmes *et al.*, *Adv. Exp. Med. Biol.*, **41A**, 43, 1974, with permission of the publisher)

and colleagues (1974). The normal intracellular concentration of purine nucleotides found in the mutant fibroblasts ruled out this mechanism (Rosenbloom *et al.*, 1968). A more likely explanation was found in the increase in the intracellular concentration of one of the substrates for the missing enzyme phosphoribosyl-1-pyrophosphate (PP-ribose-P) since it is also a substrate for the enzyme PAT. This enzyme is not saturated at the usual intracellular concentrations of PP-ribose-P. In addition Holmes *et al.* (1974) showed that PP-ribose-P converts the enzyme PAT to its more active dissociated form (Figure 13.2).

Further support for this concept has come from the discovery of an increased intracellular concentration of PP-ribose-P in fibroblasts cultured from a variety of gouty patients who all produce excessive quantities of uric acid (Becker and Seegmiller, 1974; 1975; Becker, 1976). Furthermore their cultured fibroblasts also showed an enhanced incorporation of [^{14}C]formate into FGAR in the presence of azaserine. None of these patients showed a deficiency of HPRT but in two families an enhanced activity of the enzyme PP-ribose-P synthetase was

Table 13.2   **Number of human hereditary factors recognized in various categories of inheritance**

Type of inheritance	Well documented	Presumptive
Autosomal dominant	583	635
Autosomal recessive	466	481
X-linked	93	78
Total	1142	1194
Grand Total		2236

From McKusick, 1975

demonstrably similar to that described by Zoref and colleagues (1975) but differing in the mechanism responsible for the increased activity. The PP-ribose-P synthetase of Zoref's patient failed to show a normal degree of inhibition by ADP while this enzyme in Becker's patients showed a normal responsiveness to inhibitors. In two brothers, the enzyme showed normal enzyme kinetics but a two to three-fold increase in specific activity as a dominantly inherited factor (Becker et al., 1973). PP-ribose-P synthetase of the second family showed an enhanced affinity (diminished K_m) for ribose-5-phosphate. In a third family an increased intracellular concentration of ribose-5-phosphate was found (Becker, 1976).

An additional mechanism for an increased rate of purine synthesis has come from studies of human lymphoblast lines deficient in HPRT. The rate of purine synthesis as measured by the incorporation of [^{14}C]formate into total cellular purines of normal lines can be increased to values comparable to that of HPRT deficient lines by careful elimination of all hypoxanthine from the culture medium (Hershfield and Seegmiller, 1977; Hershfield et al., 1977). This evidence suggests that hypoxanthine deprivation may play a major role in the increased purine synthesis of HPRT deficient cells since they lack the enzyme for converting this purine base to the nucleotide form.

Cells deficient in the enzyme HPRT have found a wide use in studies of cell biology and somatic cell genetics. Chemical systems have been devised for allowing the selective growth of either the mutant or the normal cell. In addition to conversion of hypoxanthine, guanine or xanthine to nucleotides, HPRT will form toxic nucleotides from either 6-thioguanine or 8-azaguanine. Consequently HPRT deficient cells but not normal cells will grow in the presence of these agents (Szybalski et al., 1962). By using this selective system on normal human fibroblasts in culture, DeMars and Held (1972) have found one to two HPRT deficient cells are formed in each cell generation of 10^6 cells thus defining the natural mutation rate at the HPRT locus.

The converse type of system for selective survival of normal cells in the presence of HPRT deficient cells is achieved by blocking purine synthesis de novo with either azaserine or amethopterin so that cells are now dependent on hypoxanthine in the media and the HPRT enzyme for providing the purine nucleotides required for survival (Fujimoto *et al.*, 1971; Littlefield, 1964). We have used the azaserine selective system to demonstrate the mutual correction of two different mutations on the X-chromosome. Fibroblasts cultured from a patient with the Mediterranean variant of glucose-6-phosphate dehydrogenase deficiency were fused with the aid of inactivated Sendai virus with HPRT-deficient fibroblasts in a nine-fold excess. Subsequent culture in azaserine killed off any unfused HPRT deficient cells but allowed the growth of clones of tetraploid cells which showed the presence of glucose-6-phosphate dehydrogenase by a histochemical method as well as HPRT by radioautographic methods. Thus both X-chromosomes remained active in the tetraploid cell (Seegmiller *et al.*, 1969).

INVESTIGATION OF THE CENTRAL NERVOUS SYSTEM

The mechanism by which HPRT deficiency leads to neurological dysfunction is not well understood. Initial suggestions that it might be the result of a toxic effect of uric acid proved unfounded. The uric acid concentration of cerebrospinal fluid of affected children was within the range found in unaffected patients. However, the concentrations of the oxypurine precursors of uric acid, hypoxanthine and xanthine were increased substantially above normal. Treatment with the xanthine oxidase inhibitor allopurinol produced a further increase in the oxypurine content of the cerebrospinal fluid but failed to exacerbate or to ameliorate the neurological symptoms to any discernible degree (Rosenbloom *et al.*, 1967). These observations favour the view that the neurological symptoms are not produced by an accumulation of oxypurines. A more likely cause of the neurological dysfunction is suggested by the discovery of a very high activity of HPRT in normal brain (Rosenbloom *et al.*, 1967) especially in the gray matter of the cerebral cortex (Adams and Harkness, 1976) suggesting that the virtual absence of HPRT activity in the brain of affected children may itself be the cause of the neurological problems. Still the mechanism connecting the enzyme defect with impaired neurological function remains obscure.

A reasonable working hypothesis would be that the enzyme defect somehow produces an imbalance in neurotransmitter activity. In favour of such a theory is the seeming over-response of affected children to emotions of excitement or anger suggestive of an excessive adrenergic response. Further evidence for such a concept is the increase in the enzyme dopamine β-hydroxylase in the plasma of all six patients who showed the full neurological symptoms of choreoathetosis, spasticity and self-mutilation. By contrast the enzyme activity was normal in the plasma of four patients who showed less severe degrees of enzyme deficiency and no neurological symptoms (Rockson *et al.*, 1974). In addition the affected children showed an inability to develop an increase in arterial pressure in response to the cold pressor test. The amelioration of the self-mutilating behaviour by administration of L-hydroxytryptophan reported by

Mizuno and Yugari in 1974 if confirmed would suggest the possibility of a relative deficit of serotonin in the central nervous system.

The inaccessibility of the central nervous system to biochemical investigation has been circumvented in part by our ability to propagate in culture a variety of cell types originating from the nervous system. Cell lines deficient in HPRT have then been selected (Wood et al., 1973; Skaper and Seegmiller, 1976a). Mouse neuroblastoma cells deficient in HPRT showed an increased intracellular concentration of the amino acid glycine, a putative neurotransmitter (Skaper and Seegmiller, 1976b). Even more interesting was a substantially diminished activity of monoamine oxidase found in HPRT deficient neuroblastoma cells (Breakefield et al., 1976) and in glioma cells (Skaper and Seegmiller, 1976c). Since the monoamine oxidase activity of glioma cells is substantially greater than that of some neuroblastoma cells the glioma cells may well have a role in the inactivation of neurotransmitter substances. In any event it provides a possible link between the HPRT deficiency and an abnormality of biochemistry involving nervous system function. In this way cell culture of these mutant cells from the nervous system has allowed us to gain further possible insight into the neurological dysfunction of this disease.

ADVANCES IN CULTURE METHODS

From the example of HPRT deficiency we can see the unique value of single gene mutations in extending our knowledge of very complicated biological processes. Complex metabolic interrelationships within the intact cell and even the intact organism that might otherwise be quite difficult or even impossible to deduce can thereby be revealed. The bacterial geneticists have long used this approach to develop at will a wide range of highly specific mutations of value in revealing the biochemical basis for ever more complex biological phenomena. Only very recently has a start been made in the use of similar approaches for the dissection of the complex biochemical interrelationships of the mammalian cell and organism. Initially the naturally occurring spontaneous mutations of patients with inborn errors were the main source available for mutant mammalian cells. A variety of selective systems has now been devised for obtaining mutant mammalian cells deficient in specific enzymes. However, for such studies the limited lifespan of cultured human fibroblasts is a serious limitation. Permanent lines of rapidly growing human cells are required. Such lines of human lymphoblasts capable of continuous growth have now been produced from the lymphocytes of peripheral blood by transformation with the Epstein-Barr virus (Sly et al., 1976). The lymphoblasts appear to have an essentially normal karyotype, grow rapidly in suspension culture with a generation time of around 18 h and can be cloned in semiliquid agar by using a feeder layer of fibroblasts with a 70–80% cloning efficiency (Lever et al., 1974, 1976; Sly et al., 1976). Using this system Dr Nuki and Dr Lever in our laboratory have been able to select a whole series of HPRT deficient lines of human lymphoblasts which show the usual increase in intracellular concentration of PP-ribose-P and an increase in activity of the first few steps of purine synthesis similar to that found in lymphoblast lines established from peripheral lymphocytes of patients with the Lesch–Nyhan syndrome (Wood et al., 1973; Lever et al., 1974). In addition we have been able to obtain lymphoblasts deficient in the enzyme adenine phosphoribosyltransferase (APRT) by culture in the purine analogue 2,6-diaminopurine (Hershfield et al., 1977). The human counterpart of the severe deficiency of this enzyme has been described in a child who excreted large amounts of adenine and 2,8-dioxyadenine in the urine (Cartier and Hamet, 1974; Simmonds et al., 1976). The latter compound had formed 'stones' in the urinary tract. Dr Hershfield in our laboratory has been able to isolate a mutant human lymphoblast line deficient in the enzyme adenosine kinase; however, the human counterpart of this defect has not yet been found in clinical medicine (Hershfield et al., 1977).

We have thus come to a new approach in the study of human hereditary metabolic defects. We are no longer dependent for such mutant human cells on the laborious clinical identification and characterization of the enzyme defect in all cases. Instead this obstacle to progress has been hurdled. We now view a new approach in which we identify and characterize human mutant cells deficient in specific enzymes all under conditions of culture in vitro. As a consequence we may very rapidly build up an inventory of enzyme defects in search of human diseases with which they may be aetiologically related. This and other developments of new systems for culture of other specialized human cell lines such as epithelial cells (Rheinwald and Green, 1975) or chondrocytes (Lust et al., 1976) give promise of further extending our knowledge of human inborn errors of metabolism at an accelerated rate.

ACKNOWLEDGEMENT

Research support for the author's laboratory is provided by grants AM-05646, AM-13622 and GM-17702 from the National Institutes of Health and by grants from the Kroc Foundation and the National Foundation. The author also acknowledges his support as a Macy Foundation Scholar at the time of this lecture.

References

Adams, A. and Harkness, R. A. (1976). Developmental changes in purine phosphoribosyltransferases in human and rat tissue. *Biochem. J.*, **160**, 565

Allison, A. C., Watts, R. W. E., Hovi, T. and Webster, A. D. B. (1975). Immunological observations on patients with Lesch–Nyhan syndrome, and on the role of *de novo* purine synthesis in lymphocyte transformation. *Lancet*, ii, 1179

Becker, M. A. (1976). Patterns of phosphoribosyl-pyrophosphate and ribose-5-phosphate concentration and generation in fibroblasts from patients with gout and purine overproduction. *J. Clin. Invest.*, **57**, 308

Becker, M. A., Kostel, P. J., Meyer, L. J. and Seegmiller, J. E. (1973). Human phosphoribosylpyrophosphate synthetase: increased enzyme specific activity in a family with gout and excessive purine synthesis. *Proc. Nat. Acad. Sci. U.S.A.*, **70**, 2749

Becker, M. A. and Seegmiller, J. E. (1974). Genetic aspects of gout. *Ann. Rev. Med.*, **25**, 15

Becker, M. A. and Seegmiller, J. E. (1975). Recent advances in the identification of enzyme abnormalities underlying excessive purine synthesis in man. *Arthritis Rheum.*, **18** (Suppl.), 687

Boyle, J. A. (1971). Gout and Pseudogout. *In*: J. A. Boyle and W. W. Buchanan (eds.) *Clinical Rheumatology*, pp. 225–226 (Oxford: Blackwell Scientific Publishers)

Breakefield, X. O., Castiglione, C. M. and Edelstein, S. B. (1976). Monoamine oxidase activity decreased in cells lacking hypoxanthine phosphoribosyltransferase activity. *Science*, **192**, 1018

Carter, M. P. and Hamet, M. (1974). A new metabolic disease: The complete deficit of adenine phosphoribosyltransferase and lithiasis of 2,8-dihydroxyadenine. *C. R. Acad. Sci.*, Paris, **279**, 883

DeMars, R. and Held, K. R. (1972). The spontaneous azaguanine-resistant mutants of diploid human fibroblasts. *Humangenetik*, **16**, 87

Fujimoto, W. Y., Subak-Sharpe, J. H. and Seegmiller, J. E. (1971). Hypoxanthine–guanine phosphoribosyltransferase deficiency: chemical agents selective for mutant or normal cultured fibroblasts in mixed and heterozygote cultures. *Proc. Nat. Acad. Sci. U.S.A.*, **68**, 1516

Garrod, A. E. (1908). The Croonian lectures on inborn errors of metabolism. Lecture II. Alcaptonuria. *Lancet*, **ii**, 73

Goldstein, J. L. and Brown, M. S. (1974). Binding and degradation of low density lipoproteins by cultured human fibroblasts. *J. Biol. Chem.*, **249**, 5153

Hershfield, M. S. and Seegmiller, J. E. (1977). Coordinate regulation of the proximal and distal steps in the pathway of purine synthesis *de novo*. *Adv. Exp. Med. Biol.* **76A**, 19

Hershfield, M., Spector, E. and Seegmiller, J. E. (1977). Purine synthesis and excretion in mutants of the WI-L2 human lymphoblastoid line deficient in adenosine kinase (AK) and adenine phosphoribosyltransferase (APRT). *Adv. Exp. Biol. Med.* **76A**, 303

Holmes, E. W., Jr., Wyngaarden, J. B. and Kelley, W. N. (1974). Human glutamine phosphoribosylpyrophosphate (PP-ribose-P) amidotransferase: kinetic, regulation and configurational changes. *Adv. Exp. Med. Biol.*, **41A**, 43

Kelley, W. N., Rosenbloom, F. M. and Seegmiller, J. E. (1967). The effect of azathioprine (Imuran) on purine synthesis in clinical disorders of purine metabolism. *J. Clin. Invest.*, **46**, 1518

Krooth, R. S. and Weinberg, A. N. (1961). Studies on cell lines developed from the tissues of patients with galactosemia. *J. Exp. Med.*, **113**, 1155

LaDu, B. N. (1972). Alcaptonuria. *In*: J. B. Stanbury, J. B. Wyngaarden and D. S. Fredrickson (eds.) *The Metabolic Basis of Inherited Disease*. 3rd edn. pp. 308–325 (New York: McGraw-Hill Book Co.)

LaDu, B. N., Zannoni, V., Laster, L. and Seegmiller, J. E. (1958). The nature of the defect in tyrosine metabolism in alcaptonuria. *J. Biol. Chem.*, **230**, 251

Lesch, M. and Nyhan, W. L. (1964). A familial disorder of uric acid metabolism and central nervous system function. *Am. J. Med.*, **36**, 561

Lever, J. E., Nuki, G. and Seegmiller, J. E. (1974). Expression of purine overproduction in a series of 8-azaguanine-resistant diploid human lymphoblast lines. *Proc. Nat. Acad. Sci. U.S.A.*, **71**, 2679

Littlefield, J. W. (1964). Selection of hybrids from matings of fibroblasts *in vitro* and their presumed recombinants. *Science*, **145**, 709

Lust, G., Nuki, G. and Seegmiller, J. E. (1976). Inorganic pyrophosphate and proteoglycan metabolism in cultured human articular chondrocytes and fibroblasts. *Arthritis Rheum.*, **19**, 479

McKusick, V. A. (1975). *Mendelian Inheritance in Man*. 4th Ed. (Baltimore: The Johns Hopkins University Press)

Mizuno, T.-I. and Yugari, Y. (1974). Self-mutilation in the Lesch–Nyhan syndrome. *Lancet*, **i**, 761

Raivio, K. O. and Seegmiller, J. E. (1972). Genetic diseases of metabolism. *Annu. Rev. Biochem.,* **41,** 543

Rheinwald, J. G. and Green, H. (1975). Serial cultivation of strains of human epidermal keratinocytes: the formation of keratinizing colonies from single cells. *Cell,* **6,** 331

Riley, I. D. (1960). Gout and cerebral palsy in a three-year-old boy. *Arch. Dis. Child.,* **35,** 293

Rockson, S., Stone, R., van der Weyden, M. and Kelley, W. N. (1974). Lesch–Nyhan syndrome: evidence for abnormal adrenergic function. *Science,* **186,** 934

Rosenbloom, F. M., Henderson, J. F., Caldwell, I. C., Kelley, W. N. and Seegmiller, J. E. (1968). Biochemical bases of accelerated purine biosynthesis *de novo* in human fibroblasts lacking hypoxanthine–guanine phosphoribosyltransferase. *J. Biol. Chem.,* **243,** 1166

Rosenbloom, F. M., Kelley, W. N., Miller, J., Henderson, J. F. and Seegmiller, J. E. (1967). Inherited disorder of purine metabolism: correlation between central nervous system dysfunction and biochemical defects. *J. Am. Med. Assoc.,* **202,** 175

Seegmiller, J. E., Rosenbloom, F. M. and Kelley, W. N. (1967). Enzyme defect associated with a sex-linked human neurological disorder and excessive purine synthesis. *Science,* **155,** 1682

Seegmiller, J. E., Siniscalco, M., Klinger, H. P., Eagle, H., Koprowski, H. and Fujimoto, W. Y. (1969). Intergenomic complementation of two X-linked genes by hybridization of mutant human fibroblasts. *Trans. Assoc. Am. Physicians,* **82,** 239

Seegmiller, J. E., Watanabe, T., Schreier, M. H. and Waldmann, T. A. (1977). Immunological aspects of purine metabolism. *Adv. Exp. Med. Biol.* **76A,** 412

Simmonds, H. A., Van Acker, K. J., Cameron, J. S. and Snedden, W. (1976). The identification of 2,8-dihydroxyadenine, a new component of urinary stones. *Biochem. J.,* **157,** 485

Skaper, S. and Seegmiller, J. E. (1976a). Hypoxanthine-guanine phosphoribosyltransferase mutant glioma cells: Diminished monoamine oxidase activity. *Science,* **194,** 1171

Skaper, S. D. and Seegmiller, J. E. (1976b). Increased concentrations of glycine in hypoxanthine–guanine phosphoribosyltransferase-deficient mouse neuroblastoma cells. *J. Neurochem.,* **26,** 689

Skaper, S. D. and Seegmiller, J. E. (1976c). Purine metabolism in thioguanine-resistant glioma cells. *Exp. Cell Res.* **100,** 415

Sly, W. S., Sekhon, G. S., Kennett, R., Bodmer, W. F. and Bodmer, J. (1976). Permanent lymphoid lines from genetically marked lymphocytes: success with lymphocytes recovered from frozen storage. *Tissue Antigens.* **7,** 165

Szybalski, W., Szybalska, E. H. and Ragni, G. (1962). Genetic studies with human cell lines. *Nat. Cancer Inst. Monogr.,* **7,** 75

Wood, A. W., Becker, M. A., Mima, J. D. and Seegmiller, J. E. (1973). Purine metabolism in normal and thioguanine resistant neuroblastoma. *Proc. Nat. Acad. Sci. U.S.A.,* **70,** 3880

Zoref, E., de Vries, A. and Sperling, O. (1975). Mutant feedback-resistant phosphoribosylpyrophosphate synthetase associated with purine overproduction and gout. Phosphoribosylpyrophosphate and purine metabolism in cultured fibroblasts. *J. Clin. Invest.,* **56,** 1093

Short Papers

Members' papers presented at the Fourteenth Symposium of The Society for the Study of Inborn Errors of Metabolism held at the University of Edinburgh on 13–16th July, 1976.

The papers will be published.

*** in full;
 ** as an abstract;
  * by title only.

*** Blaskovics, M. E., Ng, W. G. and Donnell, G. N. (Los Angeles). 'Prenatal diagnosis and a case report of isovaleric acidaemia.'

*** Divy, P., Rolland, M. O., Dingson, N., Mathieu, M., Guiband, P. and Cotte, J. (Lyons). 'Propionyl CoA carboxylase determination. Study of enzyme parameters in deficiency and control cultured skin fibroblasts.'

*** Chalmers, R. A., Lawson, A. M. and Bond, O. (London and Tromso). 'Urinary organic acids in a case of congenital lactic acidosis due to pyruvate decarboxylase deficiency.'

*** Brandt, N. J., Gregersen, N., Christensen, E. and Rasmussen, K. (Copenhagen). 'Glutaric aciduria in two brothers.'

*** Besley, G. T. N. (Edinburgh). 'Studies on sphingomyelinase activity in cultured cells and leucocytes'.

*** Maire, I., Mathieu, M., Hermier, M., Zabot, M. T. and Cotte, J. (Lyons). 'Mannosidosis. Tissue culture studies and help to prenatal diagnosis.'

 ** Kanan, M. W., Francis, M. J. O., Sykes, B. C., Ryan, T. J. and Reed, W. B. (Oxford and California). 'Epidermolysis bullosa dystrophica: abnormal numbers of 'lysosomes' in cultured fibroblasts.'

  * Niermeijer, M. F., Kleife, W. J., Heukels-Dully, M. J. and Vander Veet, E. (Rotterdam). 'Methodological aspects of prenatal diagnosis of inborn errors of metabolism.'

 ** Gitzelmann, R. and Haigis, E. (Zurich). 'Appearance of active UDP-galactose-4' epimerase in cells cultured from epimerase deficient persons.'

*** Butterworth, J. (Edinburgh). 'Diagnosis of the mucopolysaccharidoses using cultured skin fibroblasts and amniotic fluid cells.'

*** Adams, A., McVie, J. G. and Harkness, R. A. (Edinburgh). 'Adenosine deaminase activity in culture cells, lymphoid tissue and other tissues. A possible role in the immune response.'

*** Blau, K., Rattenbury, J. M., Pryse-Davies, J., Clark, P., Sandler, M., Robson, E. and Benham, F. (London). 'Prenatal diagnosis of hypophosphatasia: cytological and genetic considerations.'

  * Fowler, B., Rosenburg, L. E., Komrower, G. M. and Sardharwalla, I. B. (Yale and Manchester). 'Homocystinuria: studies of cystathionine synthase in cultured fibroblasts.'

  * Monk, A., Mitchell, A. and Holton, J. B. (Bristol). 'Activity of galactose-1-phosphate uridyl transferase in cultured cells of variose genotypes.'

 ** Duran, M. and Wadman, S. K. (Utrecht). 'Determination of optical isomers in patients with organic aciduria.'

Index